MW00784670

Virginia

Mapping the Old Dominion State through History

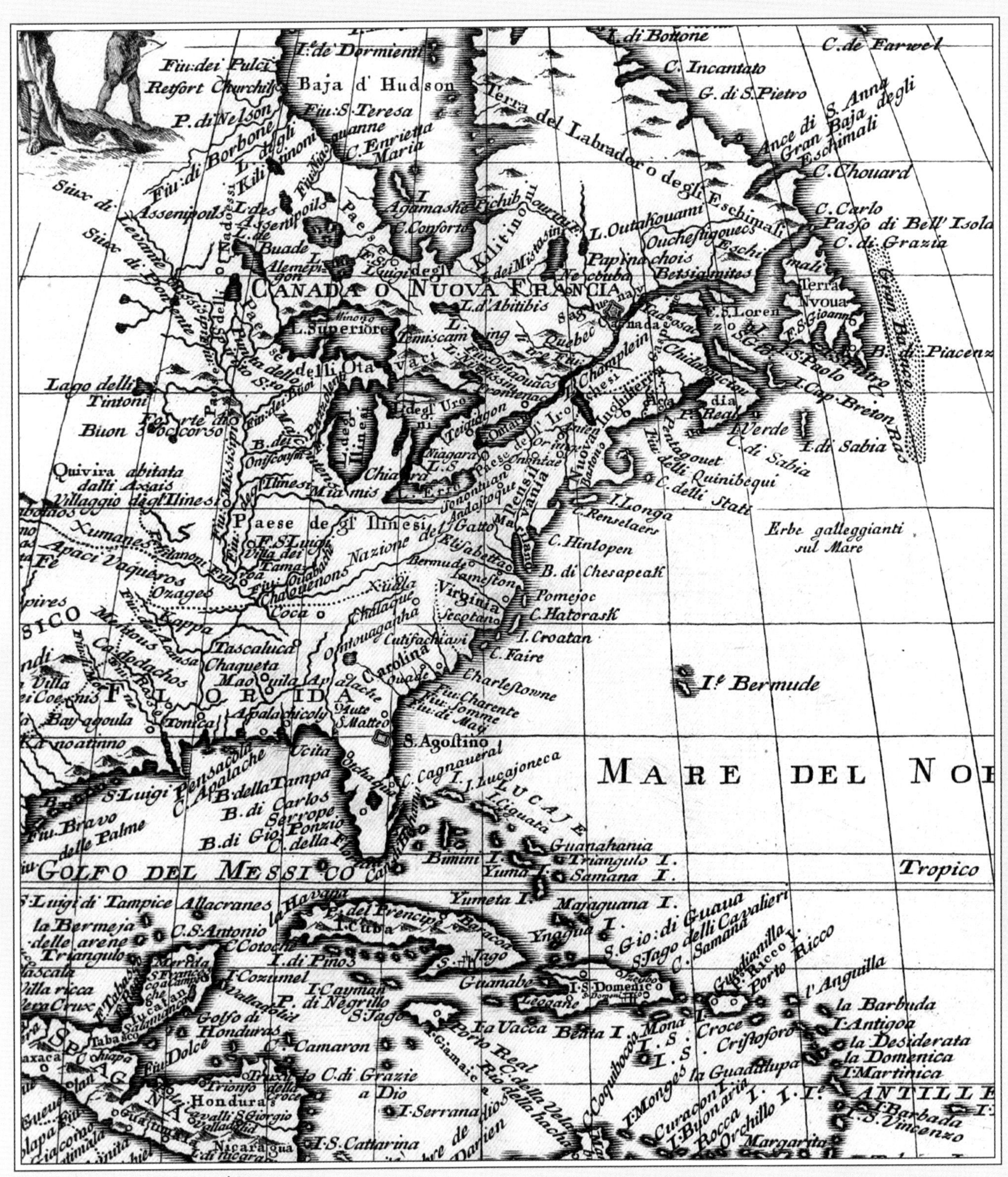

Detail of map on pages 8 and 9.

Mapping the Old Dominion State through History

Rare and Unusual Maps from the Library of Congress

Vincent Virga

and Emilee Hines

Guilford, Connecticut

Text design: Sheryl Kober
Project editor: Julie Marsh
Layout artist: Casey Shain

Library of Congress Cataloging-in-Publication Data

Virga, Vincent.
Virginia : mapping the old dominion state through history : rare and unusual maps from the Library of Congress / Vincent Virga and Emilee Hines.
p. cm.
Includes bibliographical references.
ISBN 978-0-7627-4533-3
1. Virginia—Historical geography—Maps. 2. Virginia—History—Maps. 3. Early maps—Virginia—Facsimiles. 4. Virginia—Maps, Manuscript—Facsimiles. I. Hines, Emilee. II. Library of Congress. Geography and Map Division. III. Morris Book Publishing (Firm) IV. Title.
G1291.S1V5 2010
912.755—dc22
2009018290

Printed in China

10 9 8 7 6 5 4 3 2 1

Contents

Foreword

BY VINCENT VIRGA

IN 1606 VIRGINIA'S FIRST ROYAL CHARTER WAS GIVEN in terms of latitude proving maps were close at hand. Our masterful historian Emilee Hines gorgeously equates the geography of the state with the story of its evolution and suggests Virginia may be called the Mother of Presidents *and* the Mother of States since it originally encompassed the territory of New England. The four centuries of the state's complex history unfold for us here with a stunning grace chock-full of fascinating details as varied as the book's maps because we are happily shown: "Virginia has always been in the midst of things geographically–and often historically as well."

Living on planet Earth has always raised certain questions from those of us so inclined. Of course, the most obvious one is: Where am I? Well, as Virginia Woolf sagely noted in her diary, writing things down makes them more real; this may have been a motivating factor for the Old Stone Age artists who invented the language of signs on the walls of their caves in southern France and northern Spain between 37,000 and 11,000 years ago. Picasso reportedly said, "They've invented everything," which includes the very concept of an image.

A map is an image. It makes the world more real for us and uses signs to create an essential sense of place in our imagination. (One early example of such signs are the petroglyphic maps that were inscribed in the late Iron Age on boulders high in the Valcamonica region of northern Italy.) Cartographic imaginings not only locate us on this earth but also help us invent our personal and social identities since maps embody our social order. Like the movies, maps helped create our national identity—though cinema had a vastly wider audience—and this encyclopedic series of books aims to make manifest the changing social order that invented the United States, which is why it embraces all fifty states.

Each is a precious link in the chain of events that is the history of our "great experiment," the first and enduring federal government ingeniously deriving its just powers—as John Adams proposed—from the consent of the governed. Each state has a physical presence that holds a unique place in any representation of our republic in maps. To see each one rise from the body of the continent conjures Tom Paine's excitement over the resourcefulness, the fecundity, the creative energy of our Enlightenment philosopher-founders: "We are brought at once to the point of seeing government begin, as if we had lived in the beginning of time." Just as the creators systemized not only laws but also rights in our constitution, so our maps show how their collective memory inspired the body politic to organize, codify, classify all of Nature to do their bidding with passionate preferences state by state. For they knew, as did Alexander Pope:

> All are but parts of one
> stupendous Whole
> Whose body Nature is,
> and God the soul.

And aided by the way maps under interrogation often evoke both time and space, we editors and historians have linked the reflective historical overviews of our nation's genesis to the seduction of place embedded in the art and science of cartography.

J. P. Harley posits, "The history of the map is inextricably linked to the rise of the nation-state in the modern world." The American bald eagle has been the U.S. emblem since 1782 after the Continental Congress appointed a committee in 1776 to devise an official seal for our country. The story of our own national geographical writing begins in the same period but has its roots centuries earlier, appropriately, in a flock of birds.

On October 9, 1492, after sailing westward for four weeks in an incomprehensibly vast and unknown sea during North America's migration month, an anxious Christopher Columbus spotted an unidentified flock of migrating birds flying south and signifying land—"*Tierra! Tierra!*" Changing course to align his ships with this overhead harbinger of salvation, he avoided being drawn into the northern-flowing Gulf Stream, which was waiting to be charted by Ben Franklin around the time our eagle became America as art. And so, on October 11, Columbus encountered the salubrious southern end of San Salvador. Instead of somewhere in the future New England, he came up the lee of the island's west coast to an easy and safe anchorage.

Lacking maps of the beachfront property before his eyes, he assumed himself in Asia because in his imagination there were only three parts to the known world: Europe, Asia, and Africa. To the day he died, Columbus doubted he had come upon a fourth part even though Europeans had already begun appropriating through the agency of maps what to them was a New World, a new continent. Perhaps the greatest visual statement of the general confusion that rocked the Old World as word spread of Columbus's interrupted journey to Asia is the Ruysch map of 1507. Here we see our nascent home inserted into the template created in the second century by Ptolemy, a mathematician, astrologer, and geographer of the Greco-Roman known world, the *oikoumene.*

This map changed my life. It opened my eyes to the power of a true cultural landscape. It taught me that I must always *look* at what I *see* on a map, focusing my attention on why the map was made, not who made it, when or where it was made, but *why.* The Ruysch map was made to circulate the current news. It is a quiet meditative moment in a very busy, noisy time. It is life on the cusp of a new order. And the new order is what Henry Steele Commager christened the "utopian romance" that is America. No longer were maps merely mirrors of nature for me. No longer were the old ones "incorrect" and ignorant of the "truth." No longer did they exist simply to orient me in the practical world. The Ruysch map is reality circa 1507! It is a time machine. It makes the invisible past visible.

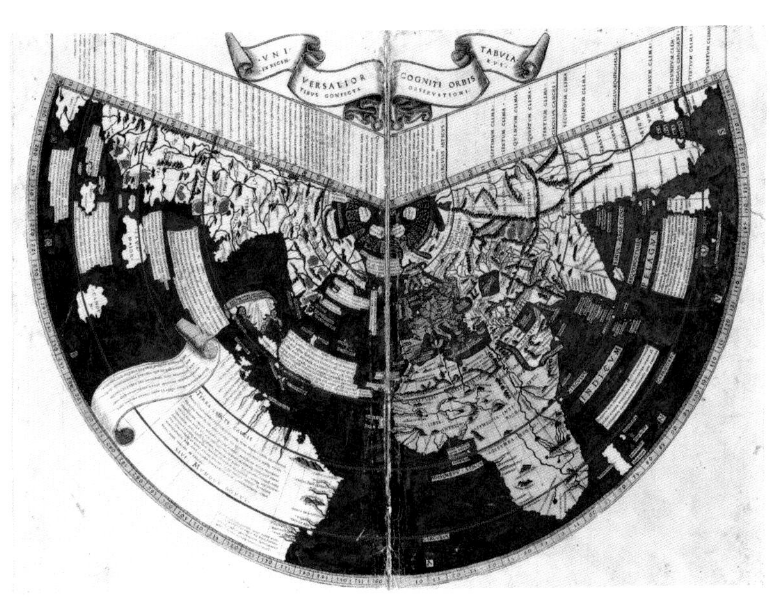

Ruysch map, 1507

Blessedly free of impossible abstractions and idealized virtues, it is undeniably my sort of primary historical document.

The same year, 1507, the Waldseemüller map appeared. It is yet another reality and one very close to the one we hold dear. There we Americans are named for the first time. And there we sit, an independent continent with oceans on both sides of us, six years *before* Balboa supposedly discovered "the other sea." There are few maps as mysterious for cartographic scholars as Waldseemüller's masterpiece. Where did all that news come from? For our purposes it is sufficient to say to the world's visual imagination, "Welcome to us Americans in all our cartographic splendor!"

Throughout my academic life, maps were never offered to me as primary historical documents. When I became a picture editor, I learned, to my amazement, that most book editors are logocentric, or "word people." Along with most historians and academics, they make their livelihood working with words and ideas. The fact of my being an "author" makes me a word person, too, of course.

But I store information visually, as does a map. (If I ask myself where my keys are, I "see" them in my mind's eye; I don't inform myself of their whereabouts in words.) So I, like everyone who reveres maps as storytellers, am both a word person and a person who can think in pictures. This is the modus operandi of a mapmaker recording the world in images for the visually literate. For a traditional historian, maps are merely archival devices dealing with scientific accuracy. They cannot "see" a map as a first-person, visual narrative crammed with very particular insights to the process of social history. However, the true nature of maps as a key player in the history of the human imagination is a cornerstone of our series.

The very title of this volume, *Virginia: Mapping the Old Dominion State through History,* makes it clear that this series has a specific agenda, as does each map. It aims to thrust us all into a new intimacy with the American experience by detailing the creative process of our nation in motion through time and space via word *and* image. It grows from the relatively recent shift in consciousness about the physical, mental, and spiritual relevance of maps in our understanding of our lives on earth. Just as each state is an integral part of the larger United States, "Where are we?" is a piece of the larger puzzle called "Who are we?"

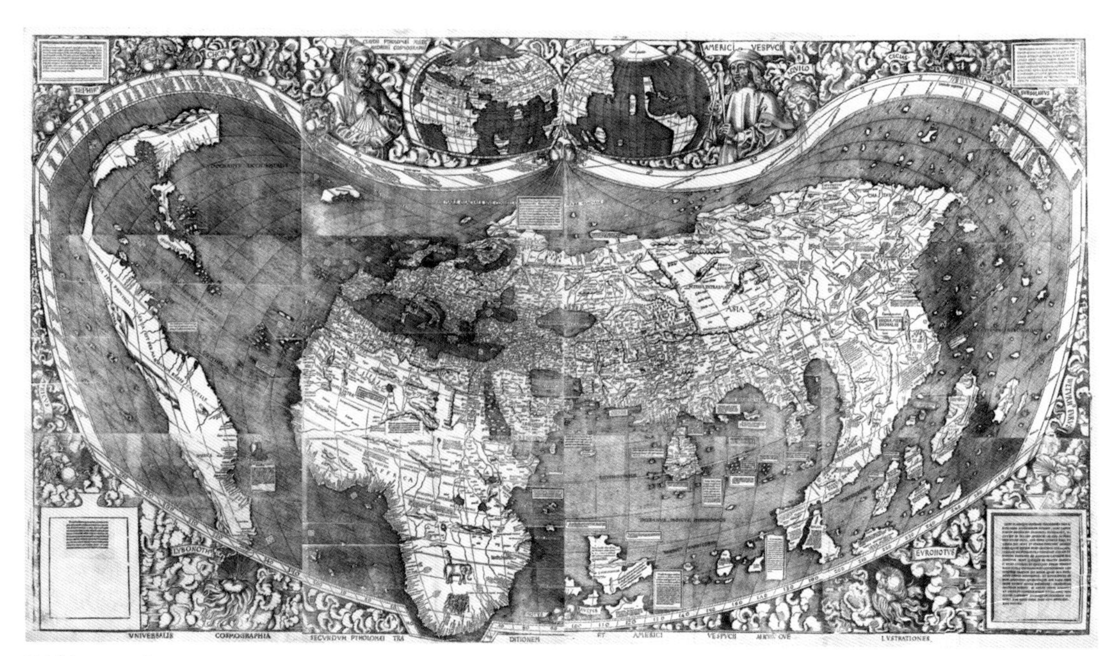

Waldseemüller map, 1507

The Library of Congress was founded in 1800 with 740 volumes and three maps. It has grown into the world's largest library and is known as "America's Memory." For me, its vast visual holdings made by those who helped build this nation make the Library the eyes of our nation as well. There are nearly five million maps in the Geography and Map Division. We have linked our series with that great collection in the hopes that its astonishing breadth will inspire us in our efforts to strike Lincoln's "mystic chords of memorys" and create living history.

On January 25, 1786, Thomas Jefferson wrote, "Our confederacy must be viewed as the nest from which all America, North and South is to be peopled." This is a man who could not live without books. This is a man who drew maps. This is a politician who in spite of his abhorrence of slavery and his respect for Native Americans took pragmatic rather than principled positions when confronted by both "issues." Nonetheless, his bold vision of an expanded American universe informs our current enterprise. There is no denying that the story of the United States has a dark side. What makes the American narrative unique is the ability we have displayed time and again to remedy our mistakes, to adjust to changing circumstances, to debate and then move on in new directions that seem better for all.

For Jefferson, whose library was the basis for the current Library of Congress after the British burned the first one during the War of 1812, and for his contemporaries, the doctrine of progress was a keystone of the Enlightenment. The maps in our books are reports on America, and all of their political programs are manifestations of progress. Our starting fresh, free of old-world hierarchies, class attitudes, and the errors of tradition, is wedded to our geographical isolation and its immunity from the endless internal European wars devastating humanity, which justify Jefferson's confessing, "I like the dreams of the future better than the history of the past." But, as the historian Michael Kammen explains, "For much of our history we have been present-minded; yet a usable past has been needed to give shape and substance to national identity." Historical maps keep the past warm with life and immediately around us. They encourage critical inquiry, curiosity, and qualms.

For me, this series of books celebrating each of our states is not about the delineation of property rights. It is a depiction of the pursuit of happiness, which is listed as one of our natural rights in the 1776 Declaration of Independence. (Thirteen years later, when the French revolutionaries drafted a Declaration of the Rights of Man, they included "property rights," and Jefferson unsuccessfully urged them to substitute "pursuit of happiness" for "property.") Jefferson also believed, "The earth belongs always to the living generation." I believe these books depict what each succeeding generation in its pursuit of happiness accomplished on this portion of the earth known as the United States. If America is a matter of an idea, then maps are an image of that idea.

I also fervently believe these books will show the states linked in the same way Lincoln saw the statement that all men are created equal as "the electric cord in that Declaration that links the hearts of patriotic and liberty-loving men together, that will link those patriotic hearts as long as the love of freedom exists in the mind of men throughout the world."

Vincent Virga
Washington, D.C.
Inauguration Day, 2009

Introduction

Located midway along the Atlantic coast of North America, about equidistant from the tropical islands of the West Indies and the icy reaches of Labrador, Virginia has always been in the midst of things geographically—and often historically as well. In springtime dogwood blossoms spray the woodlands in white, in summer lakes and the ocean beckon, in fall the hillsides blaze with color, and in winter soft snowfalls bring out skiers or those who love to see rosy sunrises out a frosty window.

Virginia is shaped like a rough triangle with a smaller triangle atop its far western tip and stretches 442 miles from east to west. The eastern side of the triangle is cut by four broad rivers—Potomac, Rappahannock, York, and James—that empty into the Chesapeake Bay. The state's topography varies from sea level to over 5,700 feet at Mt. Rogers in southwest Virginia.

Originally, Virginia encompassed territory from which all or part of seven states were formed and stretched from the Atlantic Ocean to the Great Lakes, so Virginia can be called not just the Mother of Presidents, but the Mother of States as well. Its total area now is 42,777 square miles, about a third of which is state or national park land.

Along the east coast the barrier islands shelter wild ponies and protect the mainland from the worst of Atlantic storms. Most of Virginia's Atlantic coast lies in the two counties, Northampton and Accomac, which are not joined physically to the rest of the state but are part of the Delmarva Peninsula, attached to Maryland. When Virginia was first settled, most travel was by water. With the coming of railroads and automobiles, the location of these two counties across the Chesapeake Bay presented a problem, solved for some time by ferries. Even now, ferries and tourist vessels make their way from Reedville to Tangier Island, to Maryland and to the Eastern Shore, but most cross-bay travel is by way of the Chesapeake Bay Bridge Tunnel. This engineering marvel, built in the 1950s, links the Eastern Shore counties to Norfolk. The highway soars above water, then dips beneath the bay, then up again and down again, so that two shipping lanes exist from the Atlantic into the Norfolk harbor, one of the largest natural harbors in the world.

Both the Hampton Roads and Monitor-Merrimac Tunnels, across the mouth of the James River, join Norfolk to Hampton and points north, while the Midtown and Downtown Tunnels join Portsmouth to Norfolk across the Elizabeth River. Drawbridges span narrower stretches of water in this area, called the Tidewater. From colonial times this area has been closely tied to the sea and to ships. Norfolk is home to the largest naval base in America, while the Naval Hospital, the Coast Guard Base, and the Norfolk Naval Shipyard are located in Portsmouth. Other military bases are scattered across Virginia, which was the battlefield for the Revolutionary War, the War of 1812, and the Civil War. Newport News Shipbuilding crafts both military and civilian vessels.

Along the coast of Virginia Beach, windsurfing and swimming are popular in summer, and in winter whales can be seen cavorting in the water. And all during the year tourists flock to nearby Williamsburg, Jamestown, the James River plantations, and to the many museums depicting Virginia's history and the lifestyle of her Native American inhabitants.

South of Norfolk and Portsmouth, along the border with North Carolina, lies the Dismal Swamp. In its southern portion is Lake Drummond, where cypress trees stand stark in the dark water. Despite its brown color, the water is pure and was prized by early mariners, who took it aboard their ships in barrels. The swamp gave sanctuary to runaway slaves and later was the site of hunting lodges where gentlemen could seek black bears and other prey. It has since been ditched and much of its land drained for farming and home building. The remainder, which lies within the cities of Chesapeake and Suffolk, is the Great Dismal Swamp National Wildlife Refuge. Bears still live in the Dismal Swamp and sometimes wander out, startling drivers on the busy interstate highways.

Several tidewater rivers and canals form a portion of the Inland Waterway, joining the Chesapeake Bay with North Carolina's navigable waters. George Washington long ago recognized the value of waterways to transport lumber, pitch, hides, and other goods to markets, and he invested in the area's canals.

West from the coastal cities lie the broad miles of lush tidewater farmland, where peanuts were first grown and still flourish, and where many acres now are white in autumn with bursting cotton bolls or golden with ripening soybeans. Here and on the Eastern Shore, manor houses stand at the edge of tilled fields or down a tree-lined lane.

Throughout Virginia, American Indian placenames recall the tribes of long ago, and in King William County the Mattaponi and Pamunkey, descendants of the once numerous Native Americans, have their own reservations. Every year at Thanksgiving the chief of each tribe presents the governor of Virginia with a symbolic turkey and deer, a tradition that goes back for centuries.

Northern Virginia adjoins Washington, D.C. When the nation's capital was planned, Maryland and Virginia each donated land so that the capital city would be a square, bisected by the Potomac River. However, Maryland's portion was deemed sufficient, and Virginia's donation was returned, becoming Arlington, now the site of America's largest military cemetery.

Beyond the tidewater is the fall line, the point on a river where the falls begin and beyond which ships can't pass. Richmond, Virginia's capital, developed at the fall line of the James River. In early days sailing ships brought people and supplies upstream to the river plantations and took Virginia's tobacco and other products downstream to be sent north on the Chesapeake Bay or out through the ports of Norfolk and Portsmouth to the outside world. In the nineteenth century, steamers made regular trips from Smithfield, Richmond, Hopewell, and other Virginia towns and cities to Baltimore and New York. Food grown in tidewater Virginia or oysters from the Bay could be on the dinner tables of northern cities by nightfall.

West of the fall lines, where the rivers run swift and shallow, begins the Piedmont, or "foot of the mountains," a land of gently rolling fields, sturdy redbrick houses, and white fenced pastures where horses and cattle graze. Here the soil turns from black loam to red clay, the farms are smaller and more wooded, and the views are not of marshes and the sea, but of the distant Blue Ridge Mountains. In summer the fragrance of cut grass, honeysuckle, and tobacco being cured hangs in the air. Many of the streams have been dammed to form lakes for flood control, hydroelectricity, and boating and fishing.

The tidewater planters thought of themselves as Englishmen, and they often sent their sons to England to study or hired tutors from the Mother Country. The latecomers who settled upcountry in the Piedmont were more likely to be Scotch-Irish than English and soon established colleges near their homes. These settlers thought of themselves as Virginians and were more ready than their tidewater kin to break away from England when conflict began.

Most of the early explorers and settlers came looking for gold, and one of the first Jamestown arrivals found a gold nugget washed downriver. The source was not discovered in Virginia until the 1800s, and then far upcountry from Jamestown, in Fluvanna and Buckingham Counties. Before the California Gold Rush, Piedmont Virginia was America's leading gold producer, with an output of more than $5 million. Iron ore was found in Buckingham and coal in Goochland in the 1700s, and in modern times kyanite and bauxite are mined in this region.

The land of the Piedmont slopes up to the Blue Ridge, old mountains worn by time into gently rounded mounds. This is a land of apple orchards, Christmas tree farms, ski slopes, waterfalls, and spectacular scenery, especially in autumn when the forests blaze red, orange, and brilliant yellow. The Skyline Drive runs along the crest of the Blue Ridge and a portion of the Appalachian Trail parallels it, so motorists and hikers alike can enjoy the mountains. Both are part of Shenandoah National Park. The land on both sides of the Blue Ridge south of Charlottesville, and on both sides of the Massanutten Mountains, is part of the George Washington National Forest, protected as wilderness.

Beyond the Blue Ridge lie the Shenandoah Valley, the Allegheny Mountains, and the vast Jefferson National Forest, which runs from Frederick County in the north along the West Virginia border all the way to the Kentucky border. A portion also encompasses Mt. Rogers, along the boundary with North Carolina.

This area is punctuated by underground caves and caverns, such as Luray Caverns, where visitors gaze in awe at the stalagmites and stalactites formed when water dripped through limestone. This colorful display is ever changing, for its formations are still growing.

In the far western portion of the state, coal mines have furnished the fuel needs of industry around the world for more than a century. Immigrants from central Europe dug out the black gold and stayed to become Virginians.

Virginia has everything: beaches, mountains, hiking trails, historic sites, and scenic vistas. It's a beautiful, interesting place to visit, whatever the season. As Virginians go about their daily life, they walk in the footsteps of their heroic forebears who braved wilderness, starvation, and stormy seas to make this place a home.

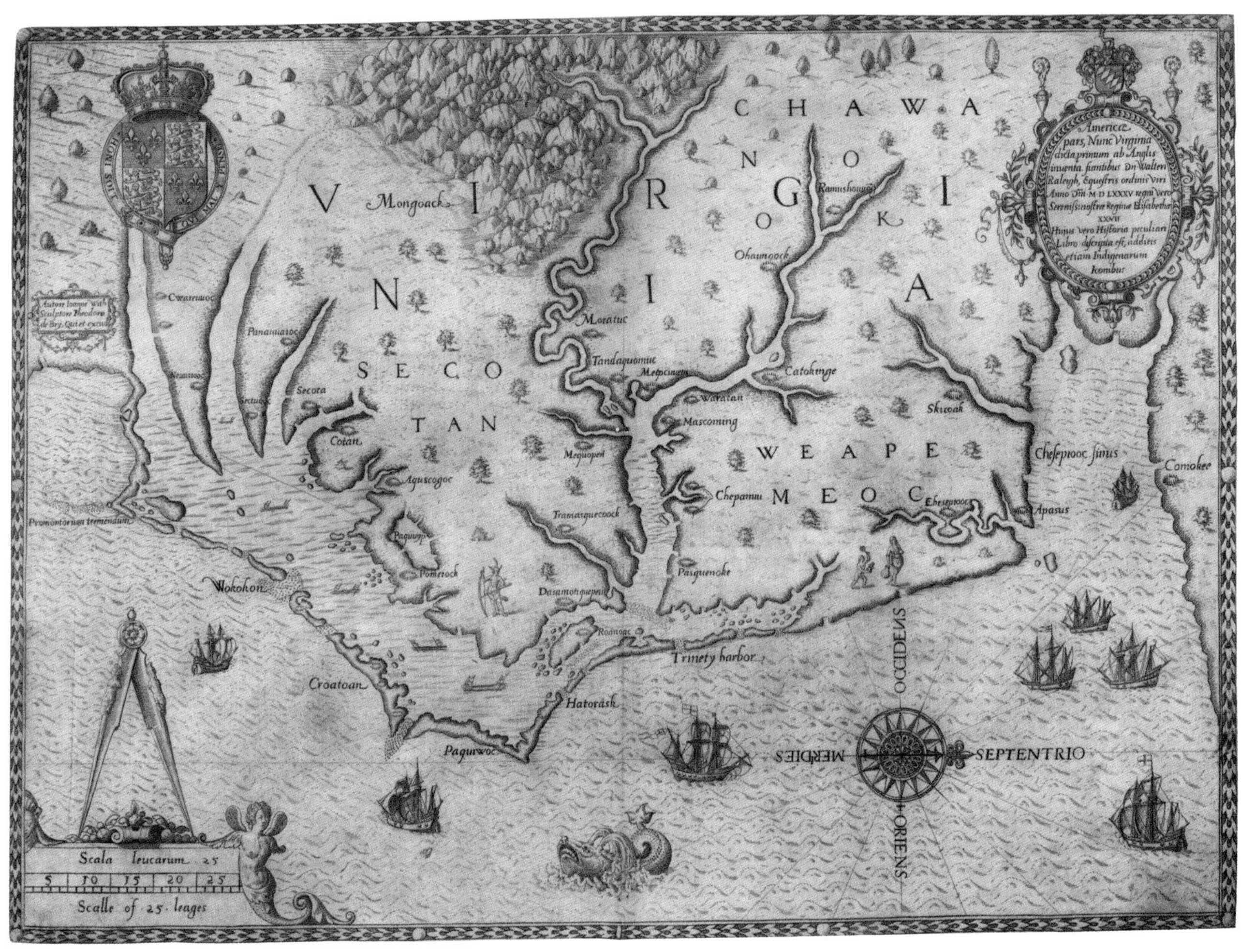

Americæ pars, nunc Virginia dicta: primum ab Anglis inuenta, sumtibus Dn. Walteri Raleigh, Equestris ordinis Viri, Anno Dni. MDLXXXV regni Vero Sereniss. nostræ Reginæ Elisabethæ XXVII, hujus vero Historia peculiari Libro descripta est, additis etiam Indigenarum Iconibus, autore Ioanne With; sculptore Theodoro de Brÿ, qui et. excud.

This area, now the coast of North Carolina, was called Virginia when the map was drawn in the late sixteenth century. Its maker, John White, was on two of Raleigh's ill-fated attempts at colonizing, in 1585 and 1588, and was the grandfather of Virginia Dare, the first English child born in America. Roanoke Island is at bottom center, and nearby is "Croatoan," where the Lost Colony may have gone. Cape Hatteras is here spelled Haterask.

Exploration and Early Settlement

VIRGINIA'S STORY BEGINS IN EUROPE. EVER SINCE Marco Polo's fabled journey overland to China, European rulers had sought a water route to the riches of the Far East. Finally, in 1492 Columbus succeeded in finding land by sailing west across the Atlantic. He soon realized that he'd found not China but a new world and claimed it all for Spain. The race for empire was on.

Five years later, Giovanni Cabotini (John Cabot) sailed for the English, exploring the coast of Canada. Next, the Portuguese king, who had sponsored expeditions along the west coast of Africa, sent Pedro Alvares Cabral westward to the New World. He touched on the coast of South America and claimed it for Portugal.

Meanwhile, the Dutch and the French sent explorers to claim a piece of the New World, and to find a water route to China around what they took to be a narrow strip of land. One mariner, Henrik (Henry) Hudson, sailed first for the Dutch, laying claim to what became New Netherland and later New York. A year later he sailed for the English, into the icy Hudson Bay, where his mutinous crew set him adrift in a small boat to a cold death.

The sixteenth century belonged to Spain. Her conquistadors conquered the Aztec of Mexico and the Inca of Peru, taking their fabulous treasure back to Spain. Spanish ships so controlled the Caribbean that it was called the Spanish Main. Spanish explorers pushed northward from Mexico into what became the western United States.

England's interest in the New World quickened when Elizabeth I came to the throne in 1558. One of her court favorites, Walter Raleigh, was determined to establish English colonies in America even though his half brother, Humphrey Gilbert, had died at sea while attempting to establish a colony in Newfoundland. Raleigh hired the writer Richard Hakluyt to publicize the planned enterprise, and in April 1584 a group set sail, landing near Cape Hatteras. They returned with glowing reports of the land the natives called Wyngandacoa. The queen said that it must instead be called Virginia in her honor as the virgin queen.

Raleigh was knighted and given claim to a vast tract of the New World, which might have made him the richest man in the world. He sent out another expedition to the same area. A group remained ashore while the ships returned to England for supplies. When Sir Francis Drake appeared unexpectedly after attacking several Spanish outposts, the explorers went aboard his ship and returned to England, taking along a plant smoked by the American Indians, tobacco. The queen found it interesting and fragrant.

The third group Raleigh sent across the Atlantic, in 1587, under the command of John White, was made up of families, including the governor's daughter, who gave birth to a daughter, Virginia Dare, soon after landing at Roanoke Island. White returned to England for supplies, not realizing that he had seen his family for the last time. When all ships were conscripted into England's war with Spain, he was unable to mount a return voyage to the New World. It was 1591 before White could return to Roanoke Island. The colony had disappeared, leaving only the word *Croatan* carved on a tree.

When Queen Elizabeth died in 1603, she was succeeded by James I, King of Scotland. James stripped Raleigh of his claims, imprisoned him, and eventually executed him. James cared nothing for Virginia and found tobacco "dangerous to the lungs."

Though the monarch was not interested in colonizing Virginia, Richard Hakluyt was, and so were English merchants. Backed by Hakluyt's persuasive writing about Virginia, a group of merchants persuaded King James to grant them a charter. Two companies were formed, the Virginia Company of London and the Plymouth Company. Their land grants overlapped between the 38th and the 41st parallels, putting them in competition. However, they were forbidden to establish colonies within one hundred miles of each other. The London Company's grant extended southward to the 34th parallel, near the present-day South Carolina border, and the Plymouth Company's grant extended north to the 45th parallel. The London Company established a colony first, and in 1609 Virginia's grant was expanded on a diagonal from the 41st parallel north westward to Lake Superior and westward to the "great sea." This was presumed to lie only a short distance west of the Atlantic.

Three small ships, *Susan Constant, Godspeed,* and *Discovery,* under the command of Capt. Christopher Newport, set sail for Virginia on December 20, 1606, crammed with food, wine, ale, muskets and gunpowder, Bibles and other religious objects, seeds, and tools. Bad weather kept them anchored in the English Channel for several weeks, but eventually they set sail across the Atlantic by the southerly route, stopping for fresh water in the Canary Islands and arriving in the Caribbean in early April. They stopped at four islands before continuing northward, arriving finally at the mouth of the Chesapeake Bay on April 26, 1607.

They went ashore, planted a cross, had a brief skirmish with Native Americans, then reboarded the ships and sailed up the broad river they named the James.

Hakluyt had written a list of instructions for the settlers. Had they been followed, the colony might have suffered less, and the death rate would undoubtedly have been lower. Hakluyt instructed them to sail upstream for one hundred miles to avoid attacks by the Spanish, and to find a place where a high river bank would make unloading the ships easier. This would have put them near present-day Richmond, not the low, marshy Jamestown. They were to fortify the place first, then explore and plant crops. To prevent anyone from running away from the colony, the sails of the pinnace (a small ship) were to be taken down and brought into the fort. The settlers were to make records of all that happened, but they were forbidden to write anything that would discourage further settlement.

The London Company sealed the list of chosen council members in a box, which was to be opened when they landed. Edward Maria Wingfield, the only investor to make the voyage to Virginia, was president of the council. A surprising member was the brave but feisty soldier, Capt. John Smith, who had spent a good part of the voyage imprisoned, accused of mutiny. He was released and soon set about mapping the area and getting to know the local Indians, the Powhatan tribe, who controlled a huge territory. By securing food from the Indians, he saved the settlers from starvation, and by his insistence on fortifying the settlement, he saved them from fatal attacks.

Captain Newport sailed for England for more supplies and food, leaving behind 104 men and boys. By the time he returned in the winter of 1608, more than 60 had died from disease—presumably malaria and typhoid, contracted from mosquitoes and from drinking the unclean, brackish river water—from Indian attacks, and from starvation. Wingfield, accused of corruption and favoritism, had been deposed as governor and kept prisoner on the pinnace, and Smith was about to be hanged, accused of murder; two of his companions, left to guard their boat while Smith was in a lodge talking with Chief Powhatan, had been killed by Indians. When Captain Newport arrived with food, supplies, and a new group of settlers, both Smith and Wingfield were exonerated, and Wingfield returned to England.

Newport explored the interior of eastern Virginia, believing Powhatan's description of "great water" just beyond the mountains, but he realized, as Smith had done, that there was no Northwest Passage through Virginia. Smith, the only remaining member of the council, was left in charge. He had a well dug within the fort, had a blockhouse built, and saw that a good crop of corn was planted. Newport arrived in August 1609 with seven shiploads of supplies and four hundred new settlers, including the first women.

Soon after, Smith was injured by a gunpowder explosion and returned to England. The Native princess Pocahontas, who Smith said had saved his life, was told that he had died. Without him, the colony floundered, and the winter that followed is known as the Starving Time, during which the colonists ate rats, dogs, and snakes, and one man was even accused of killing and eating his wife.

Early in 1610, a ship of the fleet bearing Sir Thomas Gates and other settlers to Virginia had been shipwrecked on Bermuda, inspiring Shakespeare's *The Tempest.* The group built two small ships and continued to Jamestown, arriving on May 30, 1610. They found disaster. Fewer than one hundred of the settlers remained, the settlement was in ruins, and the Native Americans had begun attacking boldly. The colonists wanted to go home. After two weeks Gates agreed and took the survivors on board his ships, abandoning Jamestown.

On the way downriver, they met three ships bringing food and supplies, 150 new settlers, and the new governor, Lord Delaware. He commanded them to return to Jamestown and set about making the colony permanent. Virginia was saved.

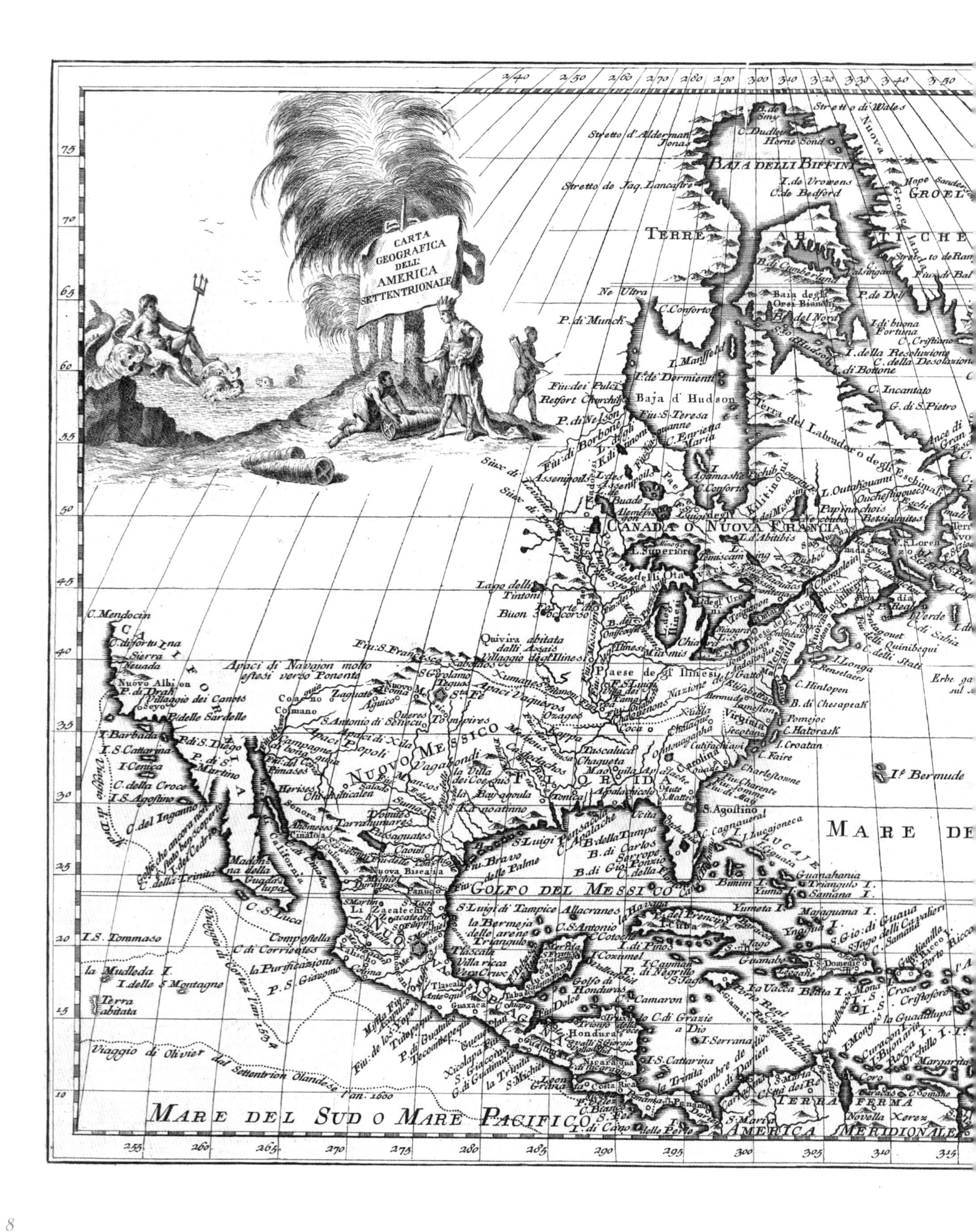
CARTA GEOGRAFICA DELL' AMERICA SETTENTRIONALE
BAJA DELLI BIFFINI
TERRE ARTICHE
Baja d' Hudson
Terra del Labrador o degli Eschimali
CANADA O NUOVA FRANCIA
L. Superiore
NUOVO MESSICO
FLORIDA
CALIFORNIA
Apaci di Navajon molto estesi verso Ponente
Quivira abitata dalli Assais
GOLFO DEL MESSICO
MARE DEL SUD O MARE PACIFICO
MARE DE
LUCAJE
I. Bermude
S. Agostino
Charlestowne
Virginia
Carolina
TERRA FERMA
AMERICA MERIDIONALE
Viaggio di Olivier dal Settentrion Olandese l'an. 1600
Stretto d' Alderman Jonas
Stretto de Jaq. Lancastre
Stretto di Wales
Ne Ultra
P. di Munck
C. Mendocin
Nuovo Albion
I. S. Tommaso
la Mudleda I.
I. delle 5 Montagne
Terra abitata
75
70
65
60
55
50
45
40
35
30
25
20
15
10
240
250
260
270
280
290
300
310
320
330
340
350
255
260
265
270
275
280
285
290
295
300
305
310
315

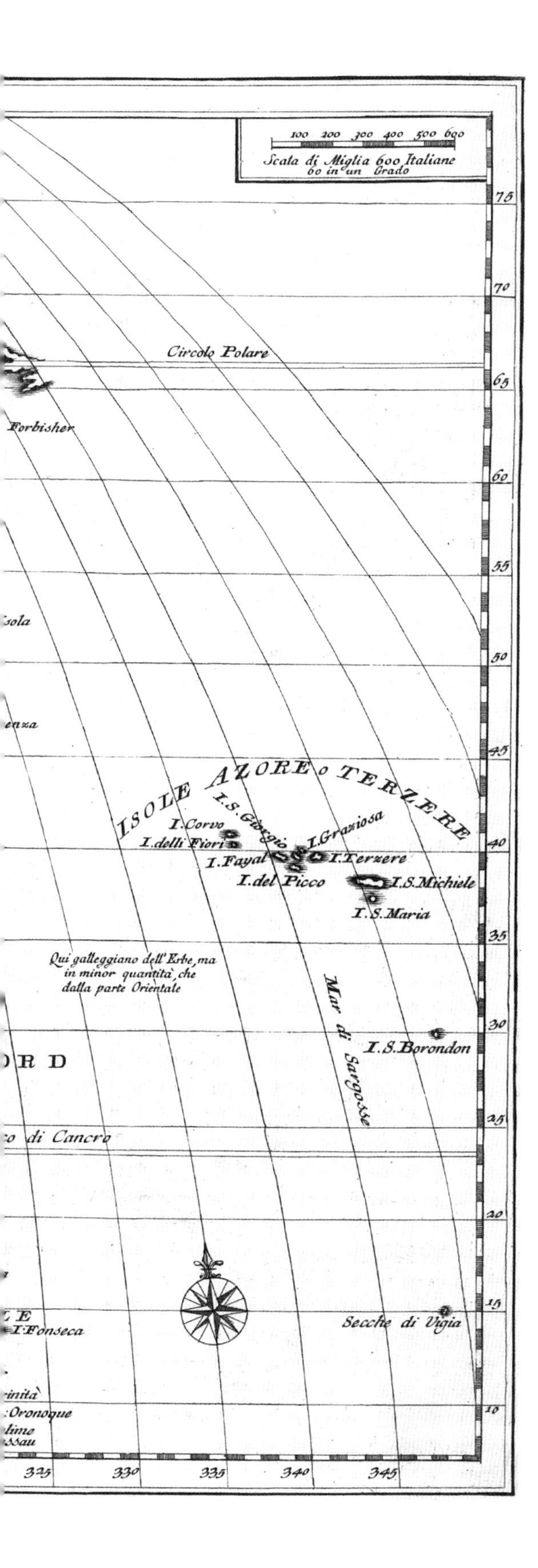

Carta geografica dell' America settentrionale.
This comprehensive map of North America, Central America, and the Caribbean, drawn by the Italian Guillaume de L'Isle, shows the extent of exploration and settlement by 1750. The Spanish empire extends from South America through Central America and far up into what became the United States. Along the east coast of North America, Virginia and the other English colonies had spread as far west as the Mississippi, and explorations had extended into Hudson Bay as mariners sought the Northwest Passage.

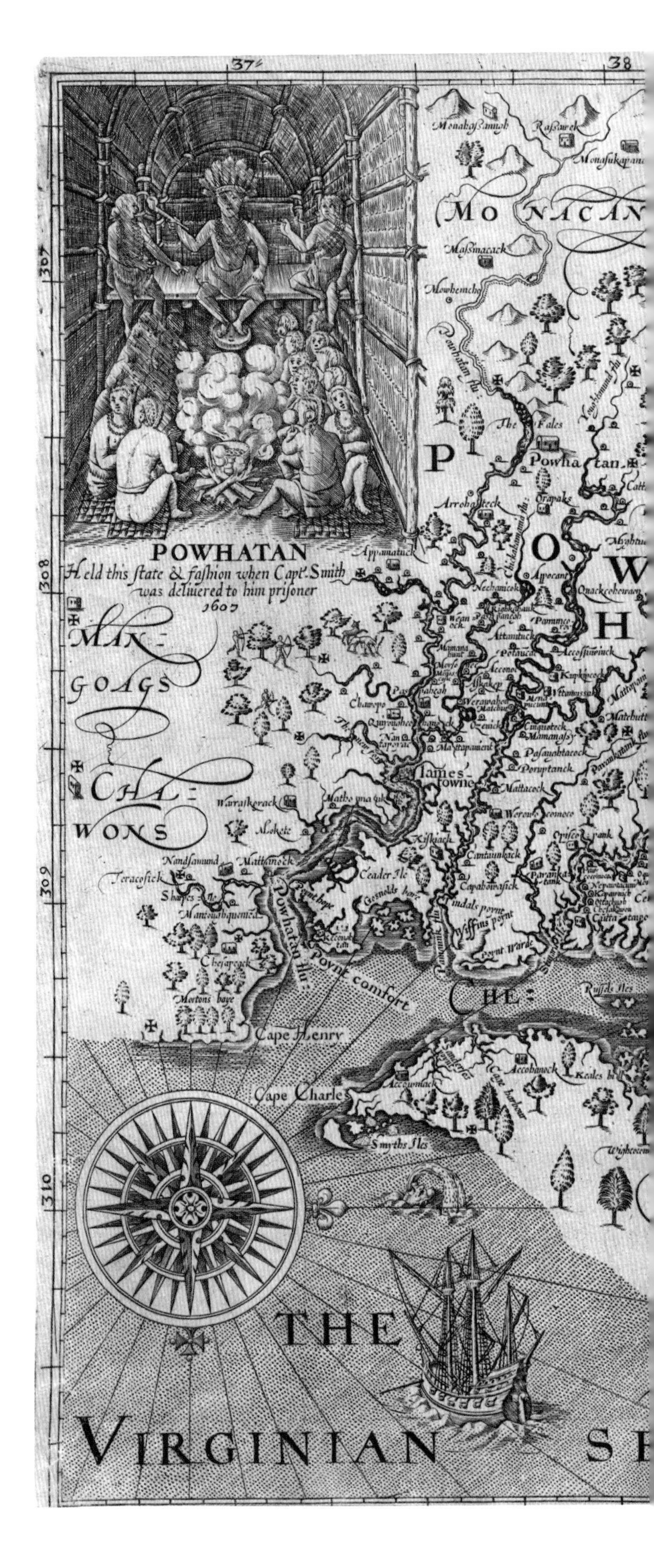

Virginia, discovered and discribed by Captayn John Smith, 1606; graven by William Hole.

Capt. John Smith was not only an able soldier and administrator, but a competent artist and cartographer, as this map of his explorations from 1607 to 1609 shows. His accurate depiction is amazing since he traveled on foot and by canoe, without modern mapping tools. He explored as far as the fall line of the James and Appomattox Rivers and located the villages and tribes that made up the Powhatans' domain. The chieftain is seen at left, a Susquehannock warrior to the right. Jamestown and Capes Henry and Charles have English names. All else, including the major river, have American Indian names.

VIRGINIA
Massaw- Massawomeck omecks
Signification of these markes,
To the crosses hath bin discouerd what beyond is by relation
Kings howses
Ordinary howses
HONI SOIT QUI MAL Y PENSE
MANN AHOACKS
Tauxsnitania
Shackaconia
Mahaskahod
Sparkes
content
Demo crites tree
Burtons
Mount
Massawteck
Cuttatawomen
Pamacocack
Namassingakent
Assaomeck
Tessamatuck
N
A
T
The Sasques-ahanougs are a Gyant like people & thus a tyred
S A S Q V E S A H A N O V G H
Tesinigh
Quadroque
Attaock
Sasquesahanough
Smyths falles
Pouels Iles
Bornes point
Winstons Iles
Brookes Forest
Ozinies
Poynt Peninsc
Tockwogh flu
Gunters Harbour
Peregrins mount
T O C K W O G H S
K V S K A R A W A O K S
A T Q V I N A C H V K E S
Chickahokin
Macocks
Scale of Leagues and halfe Leagues
Discovered and Discribed by Captayn John Smith 1606
Grauen by William Hole

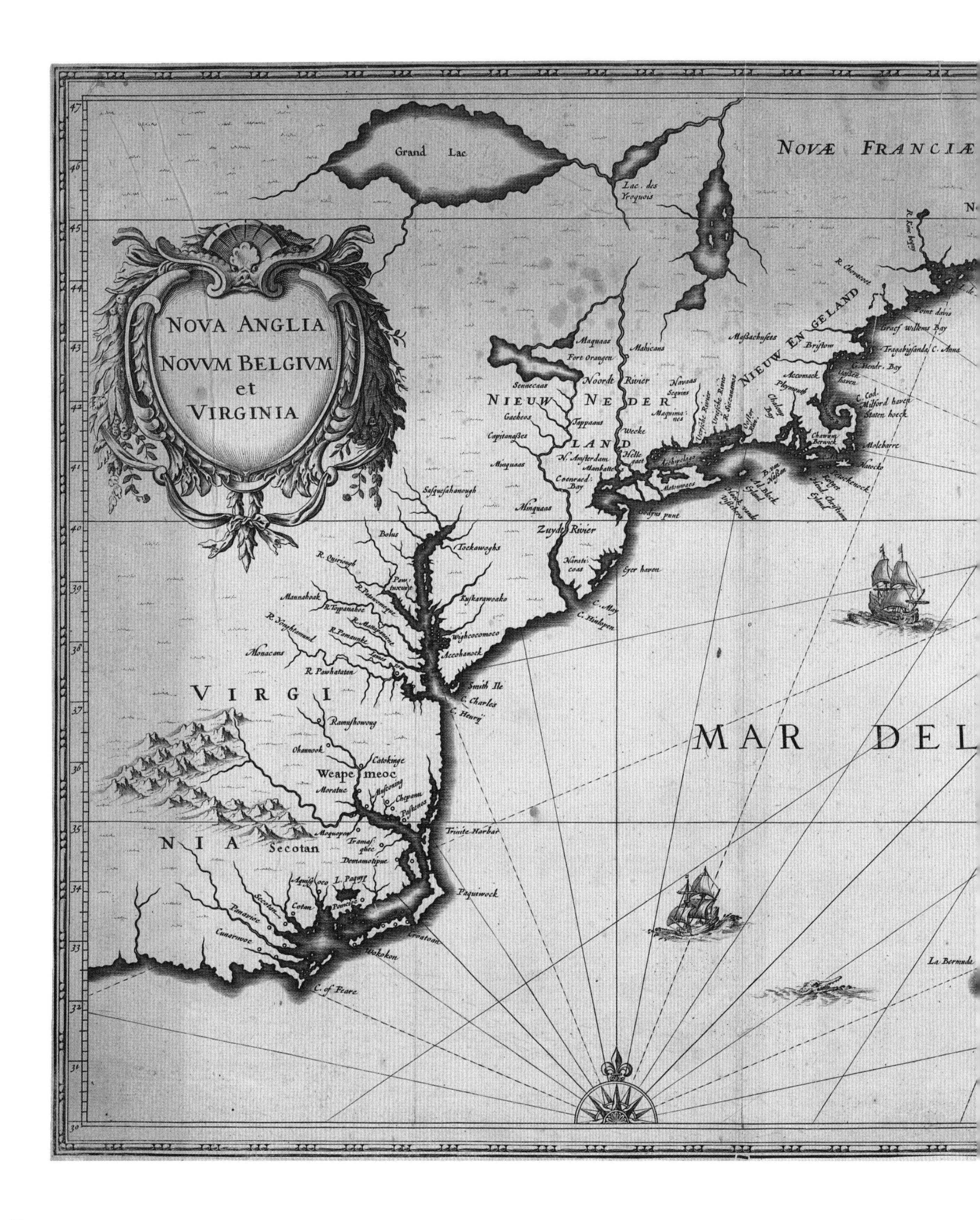
NOVA ANGLIA
NOVVM BELGIVM
et
VIRGINIA
Grand Lac
NOVÆ FRANCIÆ
NIEUW NEDER LAND
NIEUW EN GELAND
VIRGI NIA
MAR DEL

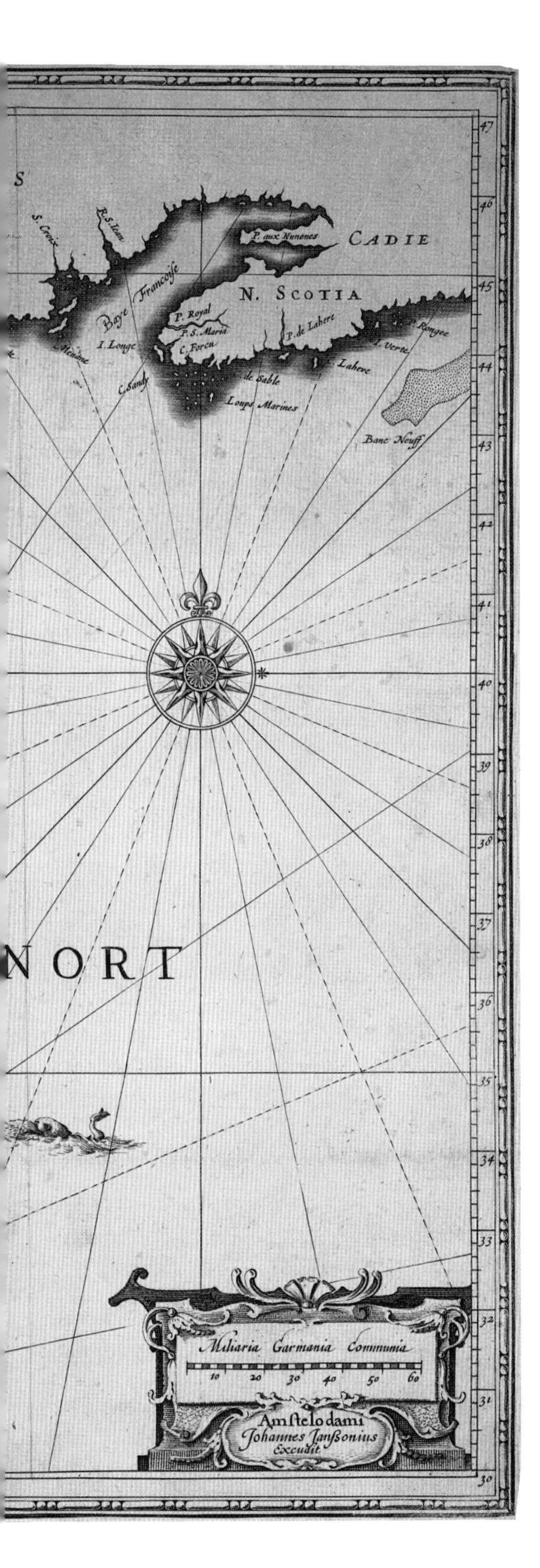

Nova Anglia, Novvm Belgivm et Virginia.

In this 1642 Jan Jansson map of New England, New Netherland, and Virginia, the mapmaker is focusing on New England. Although Nova Scotia is named and accurately drawn, there is no indication of the Maryland colony, which had been established in 1634, and the long narrow area south of Maryland, which had two Virginia counties, is much too broad. Virginia, which in that year had a population of fifteen thousand, is shown as the vast, undivided territory it still was.

Virginia and Maryland as it is planted and inhabited this present year 1670, surveyed and exactly drawne by the only labour & endeavour of Augustin Herrman bohemiensis; W. Faithorne Sculpt.

This 1670 Augustine Herrman map shows thick settlement along the major rivers of both Maryland and Virginia, especially around the fall line in Virginia, now Richmond. By 1670 the Maryland counties of Kent, Calvert, Dorchester, Anne Arundel, and Baltimore are indicated. Virginia's counties of Stafford, Middlesex, Northumberland, Lancaster, James City, Surrey, Isle of Wight, Nansemond, Norfolk, and Currituck (now North Carolina) are shown. The map is oriented with north to the right and has a portrait of Herrman, the well-known cartographer.

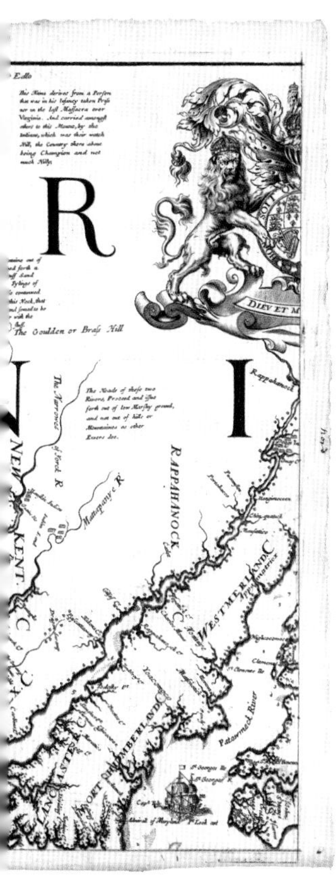
R
I
N
RAPPAHANOCK
WESTMERLAND C
NORTHUMBERLAND
The Goulden or Brass Hill

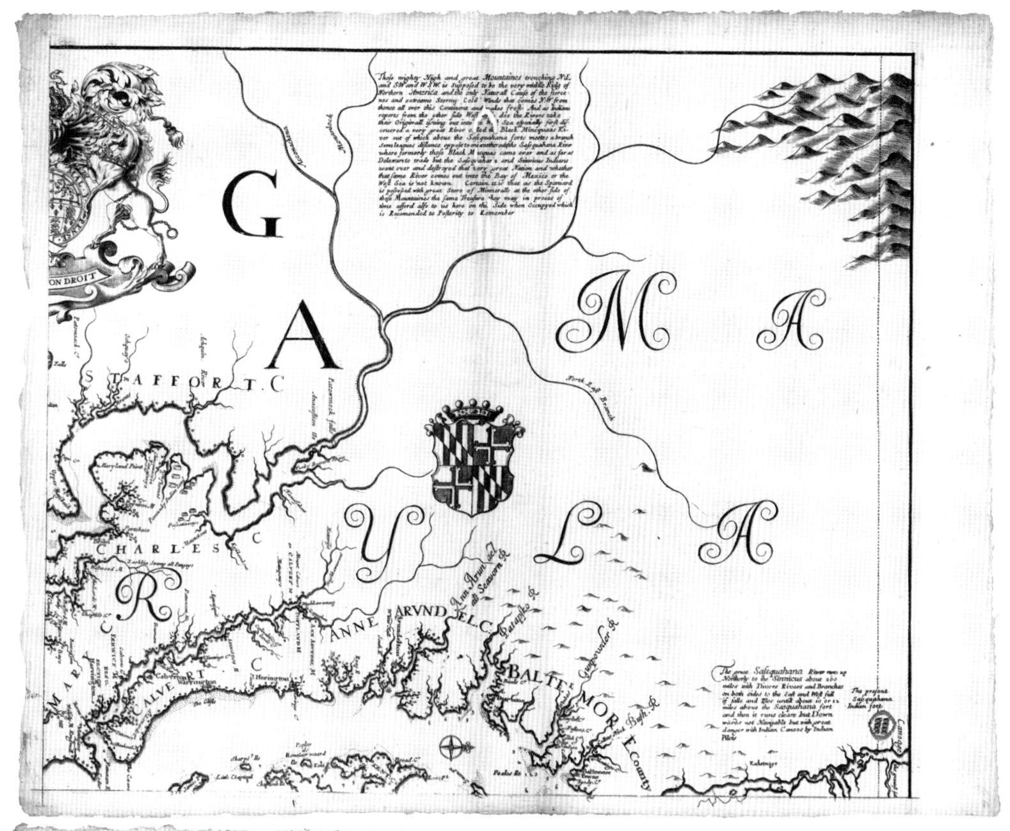
G
A
STAFFORT.C
M A
Y L A
R
CHARLES C
ANNE ARUNDEL C.
BALTEMORE County

NORTH
A Scale of 8 English Leagues

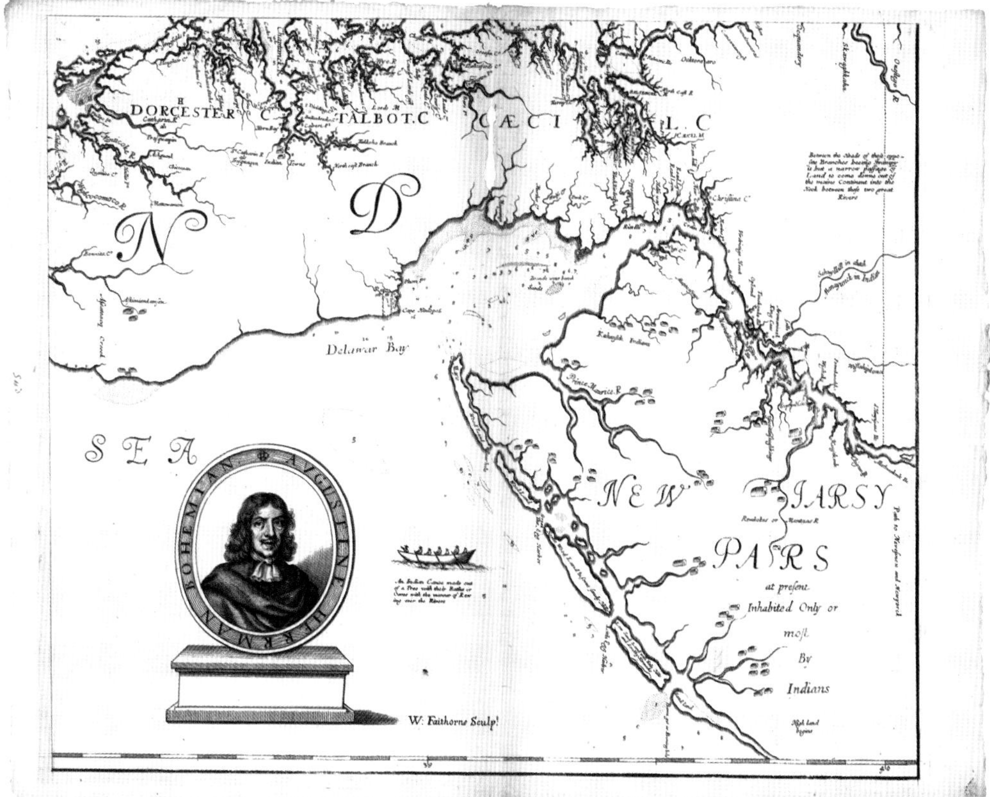
DORCESTER C
TALBOT.C
CÆCIL C
D
N
Delawar Bay
SEA
HERRMAN BOHEMIAN AUGUSTEEN
NEW JARSY
PARS
at present
Inhabited Only or
most
By
Indians
W: Faithorne Sculp:

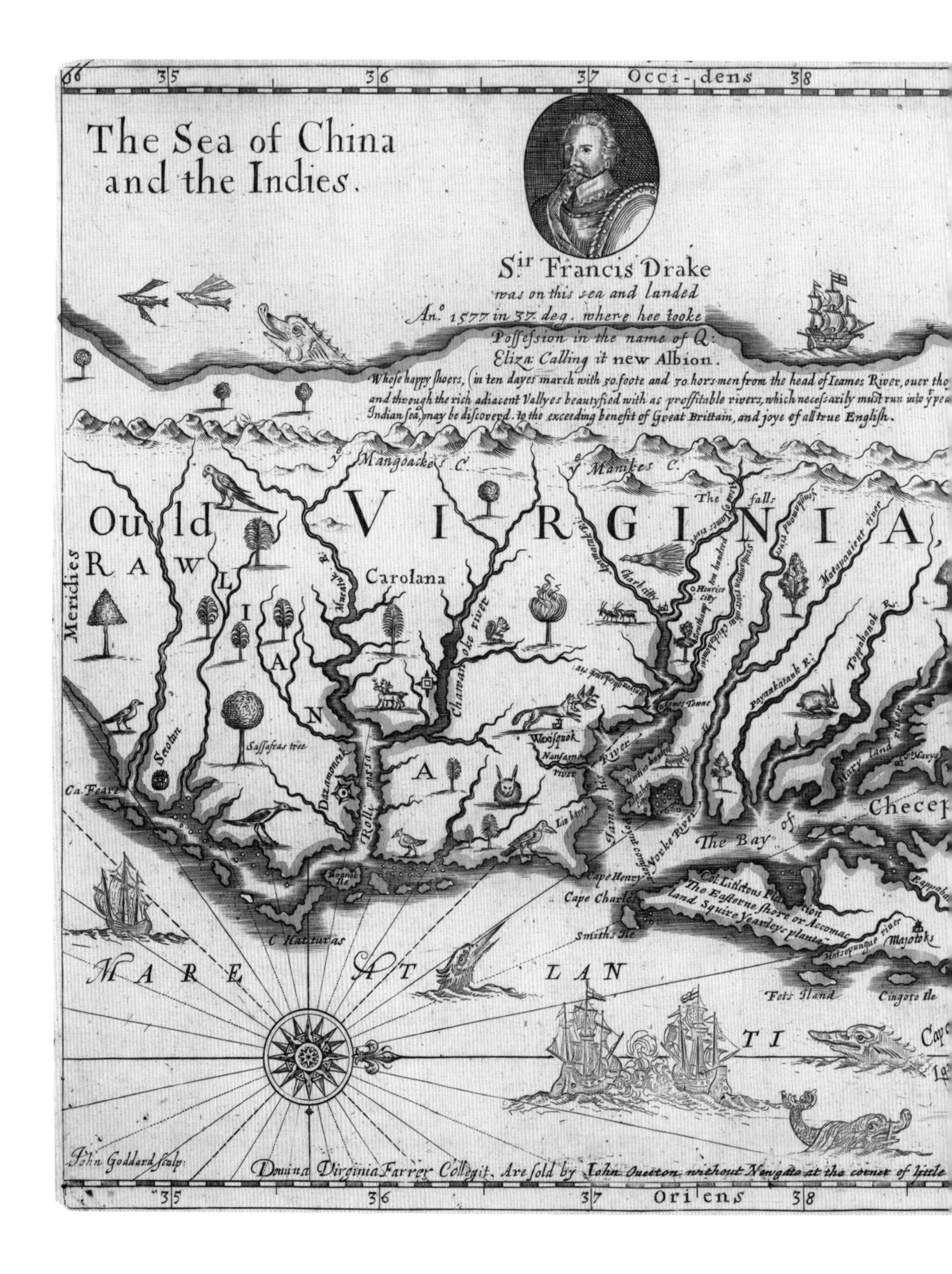

The Sea of China and the Indies.
Sr Francis Drake was on this sea and landed An.o 1577 in 37 deg. where hee tooke Possession in the name of Q: Eliza: Calling it new Albion.
Whose happy shoers, (in ten dayes march with 50 foote and 30 hors-men from the head of Iames River, ouer tho
and through the rich adiacent Vallyes beautyfied with as proffitable rivers, which necessarily must run into yͤ pea
Indian sea, may be discoverd. to the exceeding benefit of Great Brittain, and joye of all true English.
Ould
VIRGINIA
RAWLIANA
Carolana
Meridies
Chawan oke river
James Towne
Sassafras tree
Ca. Feare
Cape Henry
Cape Charles
Smiths Ile
C Hatturas
The Bay of Checep
MARE ATLANTI
Fets Iland
Cingoto Ile
John Goddard sculp:
Domina Virginia Farrer Collegit. Are sold by Iohn Overton without Newgate at the corner of little
Occi-dens
Ori ens
35
36
37
38

A mapp of Virginia discovered to ye hills, and in it's latt. from 35 deg. & 1/2 neer Florida to 41 deg. bounds of New England, Domina Virginia Farrer; John Goddard sculp. Ferrar, John, 1590?–1657.

This fanciful map, published ca. 1667 and oriented with north to the right, shows coastal Virginia and Maryland in some detail and is fairly accurate as far north as the Delaware River. However, the mapmaker, John Ferrar, still assumed that America was a narrow strip of land, and that if you proceeded beyond the mountains of Virginia, you would soon arrive at the point where Sir Francis Drake had claimed California for England. All of Canada (Nova Francia) is reduced to a small triangle, and the sea harbors strange creatures that might devour a ship.

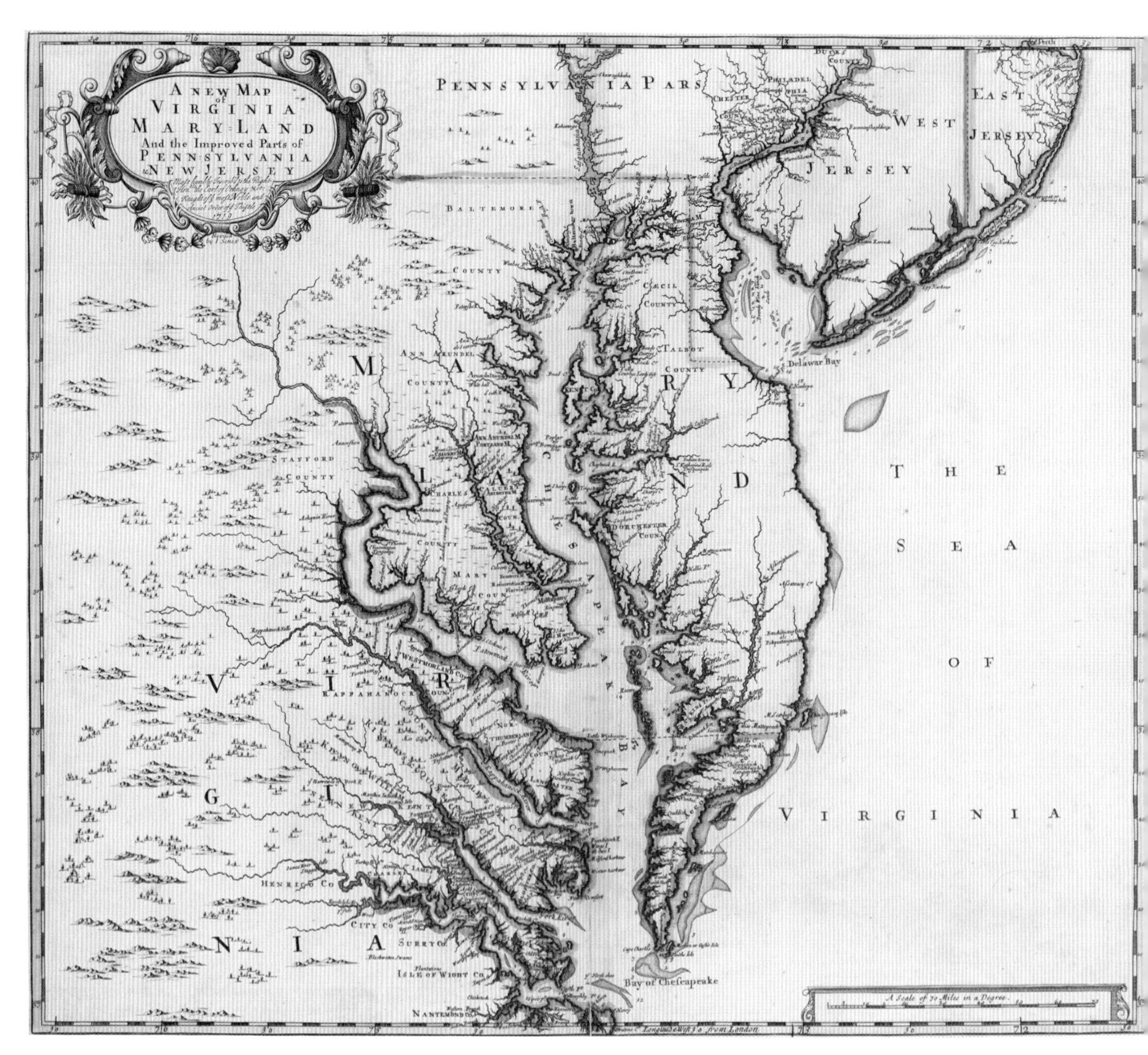

A new map of Virginia, Mary-Land, and the improved parts of Pennsylvania & New Jersey.

By 1719, when this John Senex map of Virginia, Maryland, and the improved parts of Pennsylvania and New Jersey was published, mapmaking was much more accurate than earlier depictions. The two Eastern Shore counties, Accomac and Northampton, are indicated as belonging to Virginia. Philadelphia and Baltimore have been established, and Delaware separated from New Jersey. Many Virginia places retain their American Indian names.

A Growing and Changing Colony

It took Sir Thomas Dale to set Virginia on the right track. When he arrived in Jamestown in 1611 as deputy governor, the colony was operating on communist principles: Work and food were shared. But few worked, and all ate. Dale gave the settlers land where they could grow their own food and decreed that they were to work eleven months for themselves and one month for the colony and give two bushels of corn to sustain new arrivals until their first harvest. The food supply increased dramatically, and the town grew and improved.

Dale next chose more healthful and secure places for settlement. He fortified a seven-acre peninsula in the James River near present-day Richmond and then had a ditch cut to make it an island, safe from land attack. Next, he had a series of forts built from there to the falls of the Appomattox River, near present-day Hopewell, securing another peninsula. A third fortified town, Kecoughtan, was established near the mouth of the James. Colonists could now farm safely away from Jamestown.

The London Company was impatient at the cost and lack of profit of the colony. The rumors of "gold" to be found had turned out to be false, the warm climate meant few fur-bearing animals, and winemaking and raising silkworms had failed. There had to be some way to turn a profit. The answer came from the plant the Roanoke Island explorers had taken to Queen Elizabeth: tobacco. The variety the Virginia Indians smoked was harsh, but John Rolfe, who had arrived in the colony in 1610, procured seeds of the Caribbean variety. The Virginia soil and climate were perfect to produce aromatic leaf. Soon, nearly all open land was planted in tobacco. It was popular in England and in Europe, where it was believed to cure many ailments.

This was a time of relative peace with the Powhatan tribe. Two boys, Thomas Savage and Henry Spelman, were taken to live among the Powhatans, and a young brave named Namontack came to live at Jamestown. They were to learn the language and culture of their new "family" and report back, an early "student exchange." Spelman wrote his account of Native American life and became the colony's chief interpreter.

The most famous American Indian to live among Virginians was Pocahontas, daughter of Chief Powhatan. In 1613 she was kidnapped, but her father refused the colonists' demands for ransom. John Rolfe fell in love with Pocahontas and petitioned Governor Dale for permission to marry her—as a political move that would improve the relationship between the two cultures, he said.

Although Pocahontas already had an American Indian husband, she married Rolfe in 1614 with her father's blessing. A year later their son, Thomas Rolfe, was born. The Rolfes went to England, where Pocahontas was presented at court. She died, perhaps of smallpox, as the ship bringing the family back to Virginia was leaving England. Thomas Rolfe was heir to the chieftainship of the Powhatan Indians but declined the honor.

The London Company, seeing how the colony prospered with private ownership, decided to give land as an incentive to settlers. Every colonist who had lived in Virginia for seven years received fifty acres. Fifty acres was given to each immigrant who paid his own way, and fifty acres was given to each person the immigrant brought with him. Indentured servants could work seven years for their freedom and fifty acres. This brought over four thousand new settlers between 1619 and 1625.

The London Company also decided to give Virginians a say in local matters. Each of the eleven boroughs or districts would have two representatives, who could decide on Native affairs, land patents, and other colonial concerns. At their first meeting in 1619, the representatives outlawed gambling, drunkenness, idleness, and swearing, and they required church attendance.

That year a shipload of women arrived to be wives of settlers who would pay their passage. Another cargo set Virginia on a tragic path: A Dutch ship brought Africans who were sold to Virginia planters. These Africans were given freedom and fifty acres of land after seven years, but within half a century of their arrival, Africans were enslaved.

When Powhatan died in 1618, his brother Opechancanough took control of the Powhatan Confederation. He plotted to kill the whites who were prospering, multiplying, and taking over his domain. A Christian Native, Chanco, confided the plot to his employer, who rowed frantically to Jamestown with a warning. The capital was saved, but more than three hundred settlers were killed on March 22, 1622.

Panicked, the colonists fled to the forts, where disease spread rapidly in the crowded, unsanitary conditions, and colonists died. King James blamed the London Company and in 1624 revoked their charter, making Virginia a royal colony. When he died a year later, Virginia's population had dropped to one thousand. His son Charles ignored Virginia, sending a series of royal governors. One of these, Sir William Berkeley, would dominate the colony for thirty-five years. When he arrived in 1642, Virginia's population had reached fifteen thousand, living in fifteen counties that spread from the Potomac River southward to the Dismal Swamp and westward to the fall line.

Meanwhile, the Plymouth Company had established a colony in Massachusetts and the Calvert family had begun the colony of Maryland. Virginia was not alone.

In 1644 Opechancanough struck the outlying settlements, killing five hundred colonists. He was captured and brought to Jamestown, blind and lame. Governor Berkeley intended to send him to England, but an angry soldier killed the chieftain.

Charles I was overthrown and beheaded, and his son Charles II fled from Cromwell's army. England became a commonwealth, and Loyalists flocked to Virginia. After Berkeley and the Virginia councilors swore loyalty to Charles II, Cromwell's commissioners arrived in Virginia to depose Berkeley. He retired to his country home, and Virginians got used to handling their own affairs. When Cromwell was overthrown and Charles II returned, he proclaimed his gratitude for Virginia's loyalty and called it his Old Dominion.

Charles's gratitude did not include generosity to Virginia, now a colony of forty thousand. He issued the Navigation Acts, requiring all colonial products to be sold only to England and shipped only in colonial or English vessels. This was economically devastating, especially to frontier farmers used to selling their less-desirable tobacco to the Dutch.

The frontiersmen were also being attacked by American Indians, and in the summer of 1676 they appealed to the reelected Berkeley to raise an army against the Natives. He refused, and the aggrieved settlers found a champion in twenty-nine-year-old Nathaniel Bacon. Bacon's rebels attacked the Natives and returned to threaten the governor. After several confrontations, Berkeley fled, and Bacon burned Jamestown. When Bacon died of dysentery, his army fell apart, and Berkeley took vicious revenge on the rebels. The angry king, hearing of the mismanagement of Virginians' grievances, sent a ship in 1677 to take Berkeley to England.

It was the first time Virginians had challenged royal authority, but it would not be the last.

Jamestown was rebuilt but burned again in 1698. This time the capital was moved to the middle of the peninsula, where a church and a small fort existed and a college was planned. For eighty years, the town of Williamsburg was the capital of the huge colony, which extended west to the Mississippi and north to Quebec.

Virginia was growing and changing. Germans, Scotch-Irish, and French Huguenots came, and prisoners, political and criminal, were dumped in Virginia and put to work. Younger sons of large families patented land in the Piedmont, where a speculator only needed to settle one family on each thousand acres. And there was a great deal of land. In 1670 Jacob Lederer explored Virginia and North Carolina as far as the mountains, and he saw that there was no "great sea" nearby, only more land.

After a series of quarrelsome, ineffective governors, Virginia welcomed pleasant, energetic Alexander Spotswood in 1710. He designed Bruton Parish Church, oversaw the building of much of Williamsburg, and sent out seamen who captured the pirate Blackbeard. He pacified Virginia Indians, sent troops to help South Carolina fend off Spanish attack, and with the governor of North Carolina settled the Virginia-Carolina boundary.

Spotswood knew that Virginia's future lay to the west. He led explorers across the Blue Ridge into the Shenandoah Valley, calling his group the Knights of the Golden Horseshoe. Soon the Valley would fill with settlers, many coming from Pennsylvania. As settlement moved west, new counties were formed: By 1727, when Governor William Gooch arrived, Virginia had one hundred thousand people living in twenty-eight counties. At the end of his term twenty-two years later, Virginia's forty-four counties included Kentucky, Tennessee, and Ohio Counties, which each became states, and Greenbriar County, which brought English settlement into contact and conflict with the French.

After Spotswood was forced out by Virginia planters who feared losing their privileges, he remained in Virginia, set up an ironworks as an alternate to tobacco growing, became postmaster general of the colonies, and lived in the county named for him, Spotsylvania. Had all colonial governors been as wise, revolution might have been avoided.

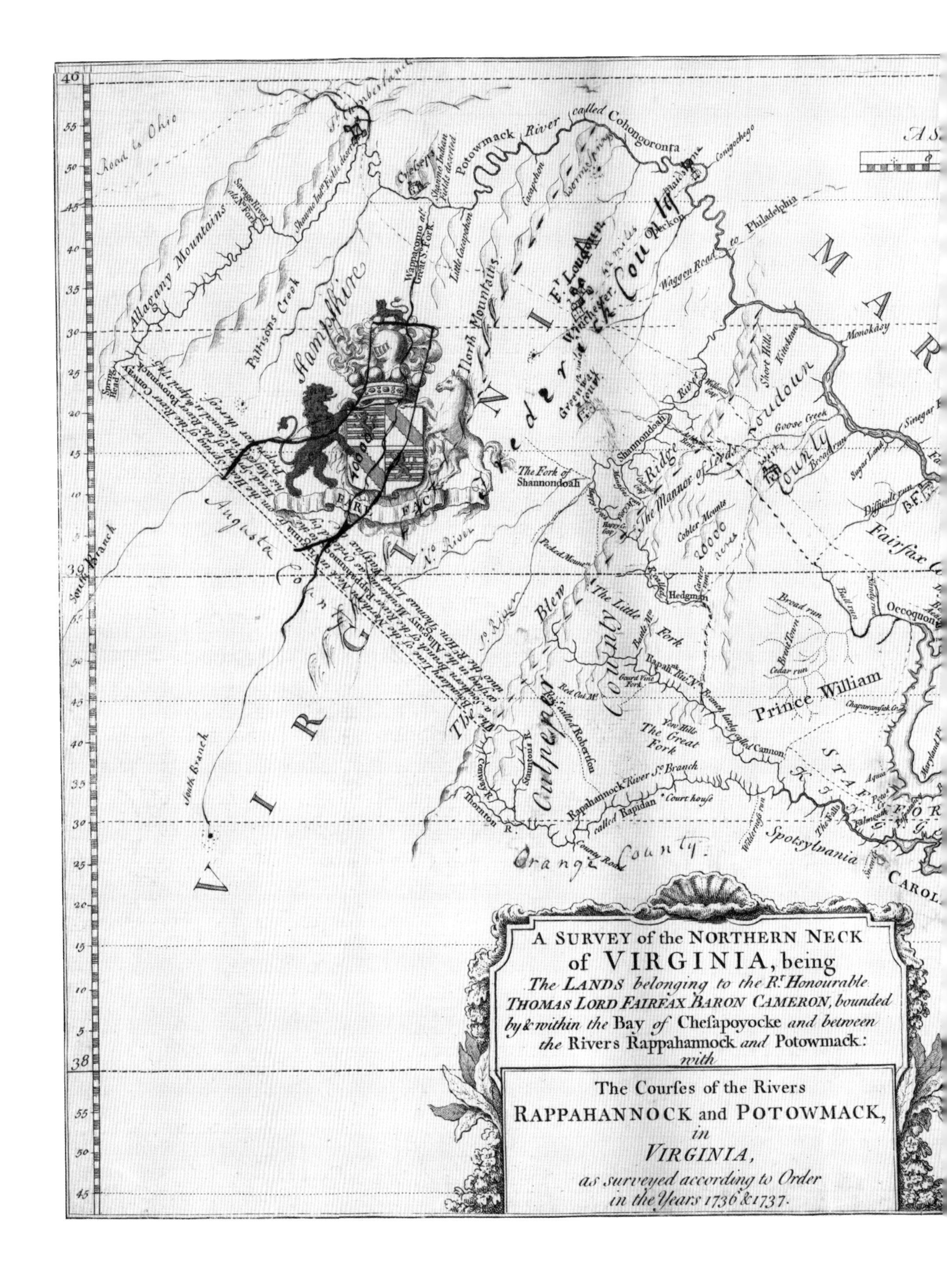
A SURVEY of the NORTHERN NECK of VIRGINIA, being The LANDS belonging to the R.t Honourable THOMAS LORD FAIRFAX BARON CAMERON, bounded by & within the Bay of Chesapoyocke and between the Rivers Rappahannock and Potowmack: with
The Courses of the Rivers RAPPAHANNOCK and POTOWMACK, in VIRGINIA, as surveyed according to Order in the Years 1736 & 1737.
Potowmack River called Cohongoronta
Road to Ohio
Allagany Mountains
Hampshire
North Mountains
Waggon Road to Philadelphia
The Fork of Shannondoah
Shannondoah
The Manor of Leeds
Loudoun County
Goose Creek
Fairfax
Prince William
The Little Fork
The Great Fork
Rappahannock River So. Branch called Rapidan
Court house
Orange County
Spotsylvania
South Branch
Thornton R.
Cobler Mounts
Hedgman
Monokasy
Occoquon
Augusta County
VIRGINIA
MARYLAND

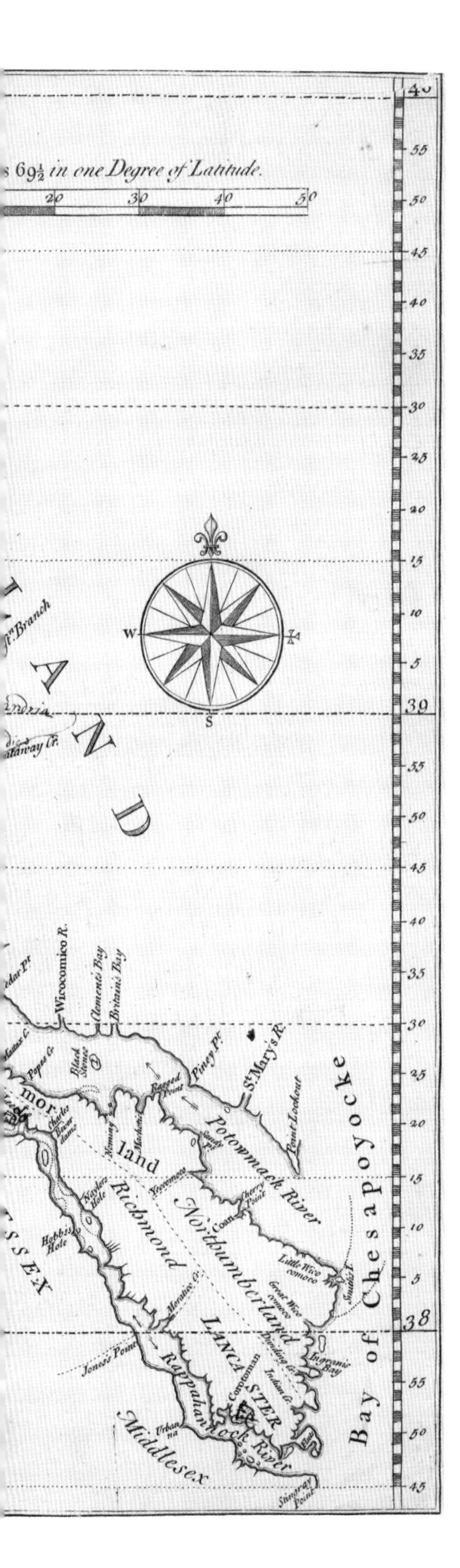

A survey of the northern neck of Virginia, being the lands belonging to the Rt. Honourable Thomas Lord Fairfax Baron Cameron, bounded by & within the Bay of Chesapoyocke and between the rivers Rappahannock and Potowmack: With the courses of the rivers Rappahannock and Potowmack, in Virginia, as surveyed according to order in the years 1736 & 1737.

This map by John Warner shows the northern neck of Virginia, surveyed 1736–1737, indicating the six million acres belonging to Thomas Lord Fairfax, with the "Bay of Chesapoyocke" as the eastern boundary and between the Rappahannock and "Potowmack" Rivers. His home, Belvoir, is indicated near the town of Occoquan. Also indicated on the map are the road to Ohio and the Wagon Road to Philadelphia.

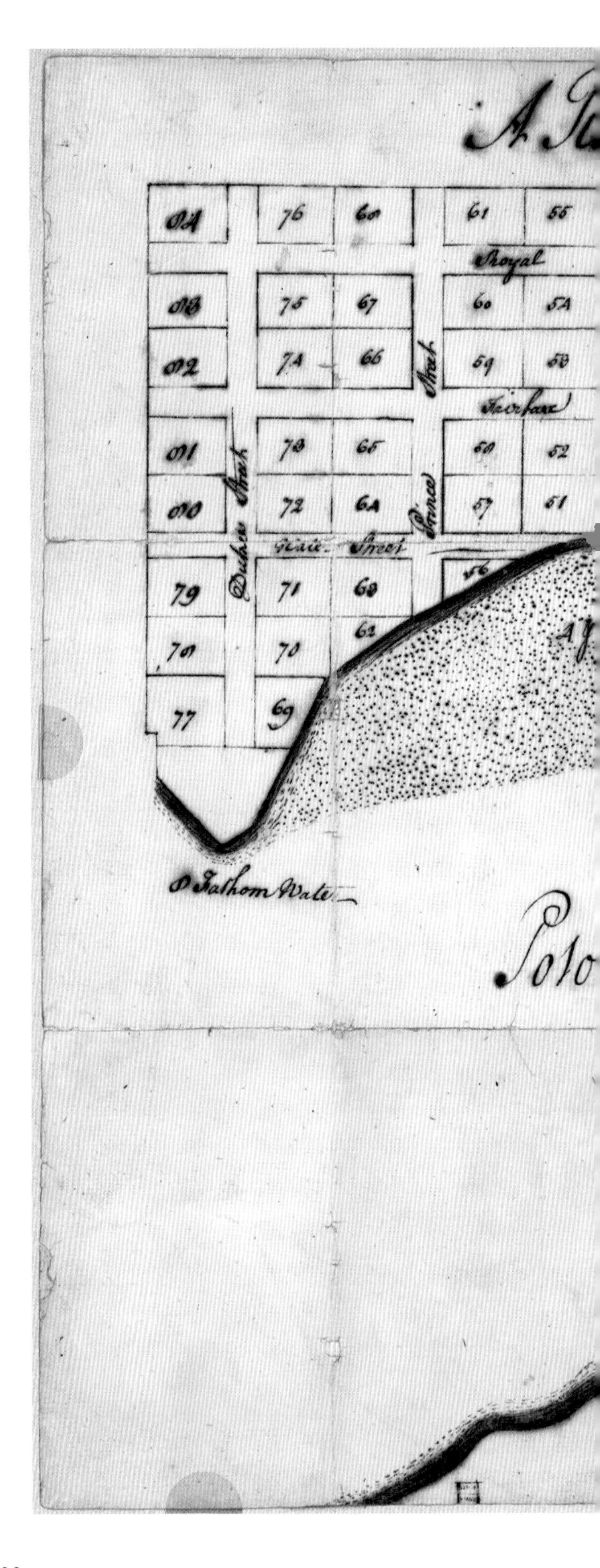

A plan of Alexandria, now Belhaven. Washington, George 1732–1799.

A map of lots in the town of Alexandria, drawn by George Washington in 1749, when he was only seventeen. Alexandria was established in 1748 as a trading center for Mount Vernon and Gunston Hall. Owners of the lots included Lawrence Washington, Augustine Washington, William Fairfax, Col. George Fairfax, and, surprisingly for that time, a woman, Anne West.

f Alexandria now Belhaven
Cameron Street
Queens Street
Water Street
Fathom Water
k River
Maryland
Proprietors Names
Col. W. Fitzhugh
Harry Piper
Roger Lindon
Jn. Dalton
Allan McCrae
John Carlyle
Wm. Ramsay
Law. Washington
Hon. Wm. Fairfax
Col. Geo. Fairfax
Col. Nath. Harrison
Nath. Chapman
Gerrd. Alexander
John Alexander
John Dalton
Henry Fitzhugh
Hugh West
John Pagan
Ralph Wormeley Esq.
George Mason
William Ramsay
Col. W. Fitzhugh
John Peyton
John West Senr.
augustine Washington
anne West
Wm. Henry Terrett
Pearson Terrett
John Champe
George West
Hugh West Junr.
Wm. West Junr.
774

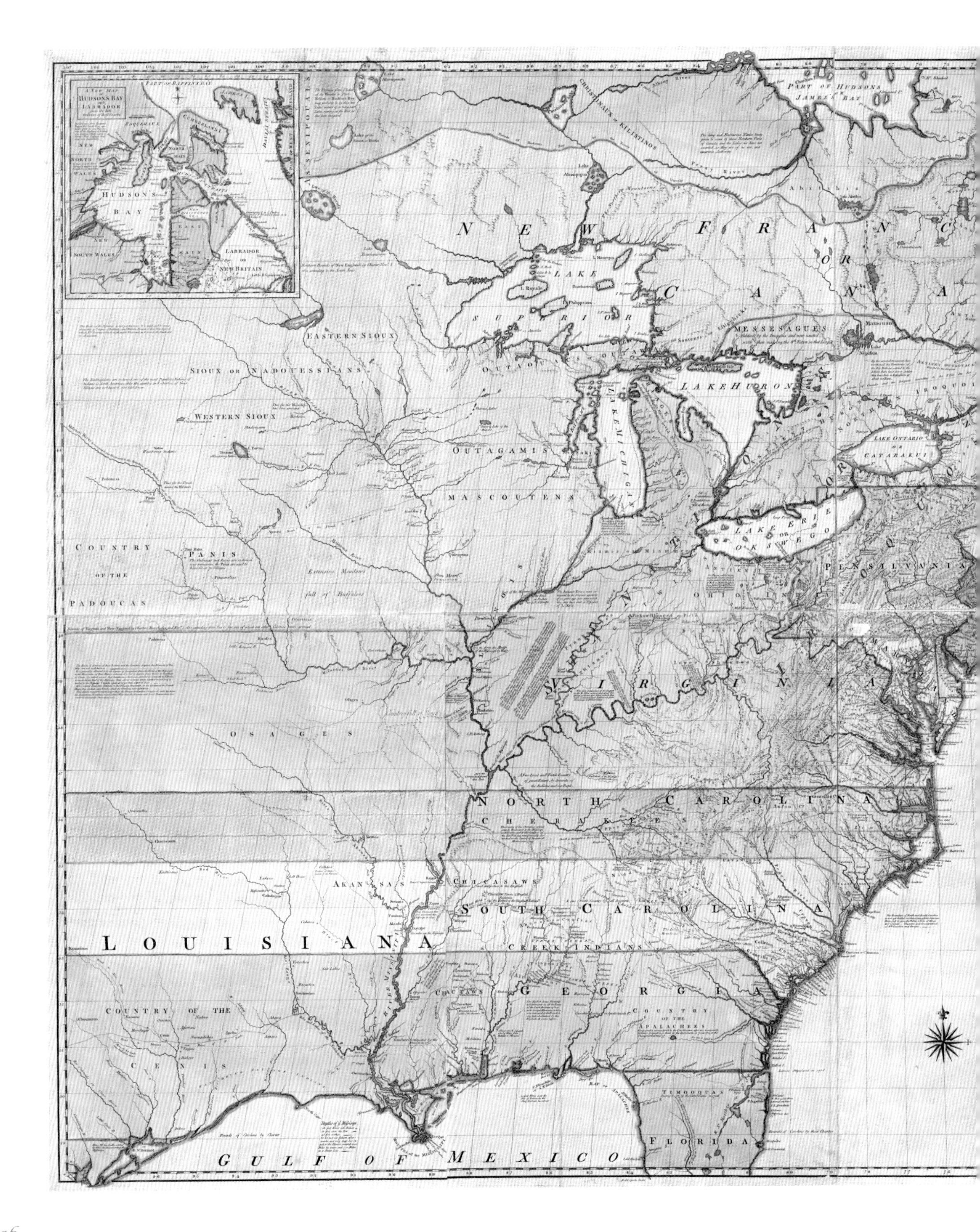

A New Map of Hudsons Bay and Labrador
Part of Baffins Bay
Hudsons Bay
Labrador or New Britain
New South Wales
New North Wales
Cumberland
Part of Hudsons James's Bay
N E W F R A N C E
C A N A D A
Lake Superior
Lake Huron
Lake Michigan
Lake Erie or Oswego
Lake Ontario or Catarakui
Eastern Sioux
Sioux or Nadouessians
Western Sioux
Outagamis
Mascoutens
Messesagues
Christinaux or Kilistinos
Country of the Padoucas
Panis
Extensive Meadows full of Buffaloes
Osages
Virginia
Pensalvania
Ohio
North Carolina
Cherakees
South Carolina
Chicasaws
Akansas
Louisiana
Creek Indians
Georgia
Chactaws
Country of the Apalaches
Country of the Cenis
Timooquas
Florida
Gulf of Mexico
River Mississippi

A map of the British and French dominions in North America, with the roads, distances, limits, and extent of the settlements, humbly inscribed to the Right Honourable the Earl of Halifax, and the other Right Honourable the Lords Commissioners for Trade & Plantations, by their Lordships most obliged and very humble servant, Jno. Mitchell. Tho: Kitchin, sculp.

This map, drawn by John Mitchell and published in 1755, shows the British and French claims in North America, with roads, distances, limits, and extent of settlements indicated. Virginia's borders with North Carolina and Maryland had been determined, but the colony's westward claims extended west to the Pacific Ocean and northward to the Great Lakes. The Ohio Valley of Virginia would soon be the site of the French and Indian War.

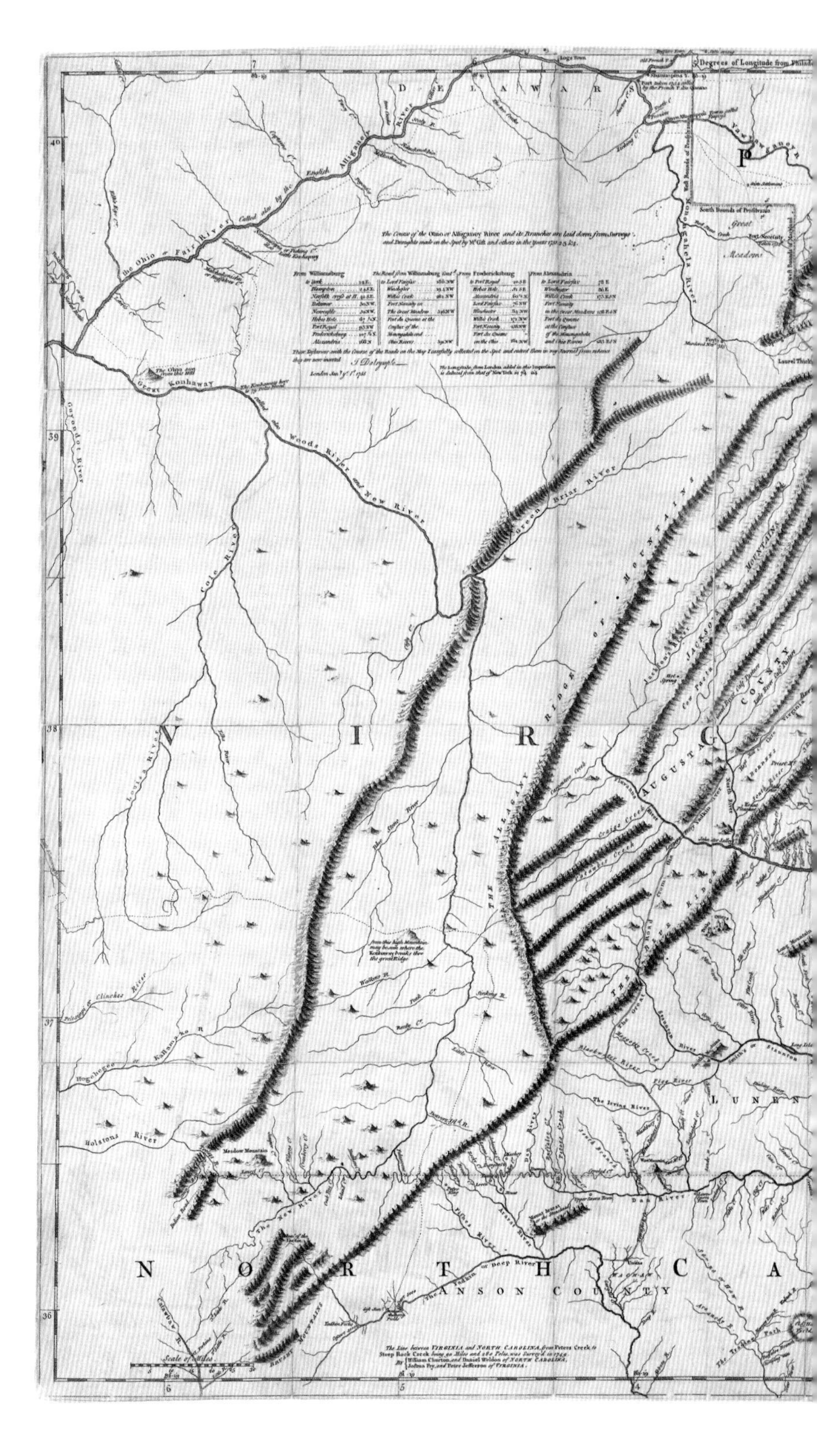

A map of the most inhabited part of Virginia containing the whole province of Maryland with part of Pensilvania, New Jersey and North Carolina. Drawn by Joshua Fry & Peter Jefferson in 1751.
This map, drawn in 1751 by Joshua Fry and Peter Jefferson, shows "the most inhabited part of Virginia containing the whole province of Maryland with part of Pensilvania, New Jersey and North Carolina." Ferry routes and the Dismal Swamp are indicated, as are the chains of mountains in western Virginia. The eastern portion of the state had been settled and divided into counties, and settlement would soon leap the mountains and expand into the Ohio Valley.

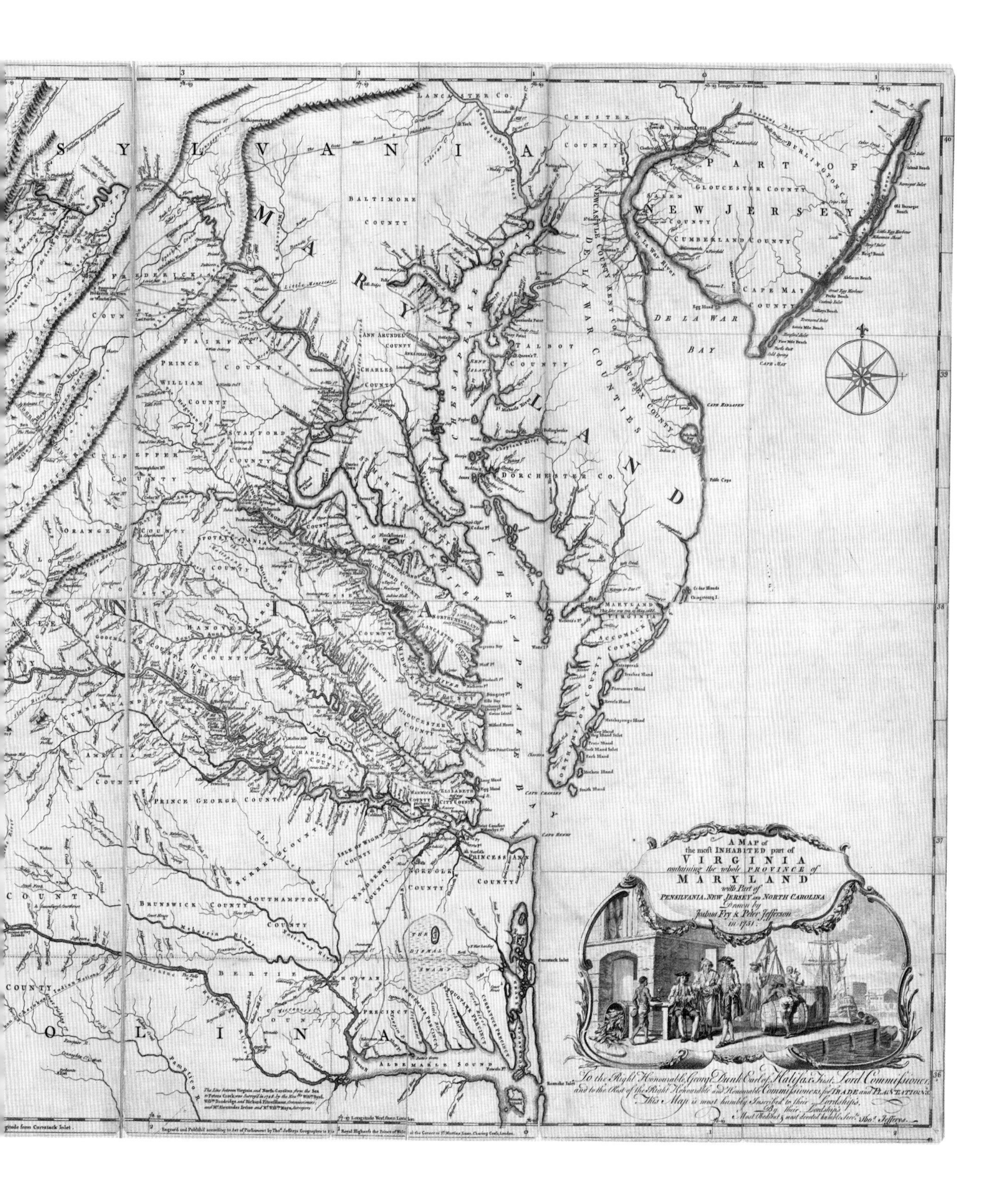
A MAP of
the most INHABITED part of
VIRGINIA
containing the whole PROVINCE of
MARYLAND
with Part of
PENSILVANIA, NEW JERSEY and NORTH CAROLINA
Drawn by
Joshua Fry & Peter Jefferson
in 1751.
To the Right Honourable George Dunk Earl of Halifax First Lord Commissioner;
and to the Rest of the Right Honourable and Honourable Commissioners for TRADE and PLANTATIONS;
This Map is most humbly Inscribed to their Lordships,
By their Lordships
Most Obedient & most devoted humble Servt. Thos. Jefferys.
PART OF
NEW JERSEY
GLOUCESTER COUNTY
CUMBERLAND COUNTY
CAPE MAY COUNTY
DE LA WAR BAY
BALTIMORE COUNTY
ANN ARUNDEL COUNTY
CHARLES COUNTY
TALBOT COUNTY
DORCHESTER CO.
ACCOMACK COUNTY
PRINCE WILLIAM COUNTY
STAFFORD COUNTY
ORANGE COUNTY
HANOVER COUNTY
GLOUCESTER COUNTY
PRINCE GEORGE COUNTY
BRUNSWICK COUNTY
SOUTHAMPTON COUNTY
NORFOLK COUNTY
PRINCESS ANN COUNTY
ALBEMARLE SOUND
CHESTER COUNTY
LANCASTER CO.
CAPE MAY
CAPE CHARLES
CAPE HENRY

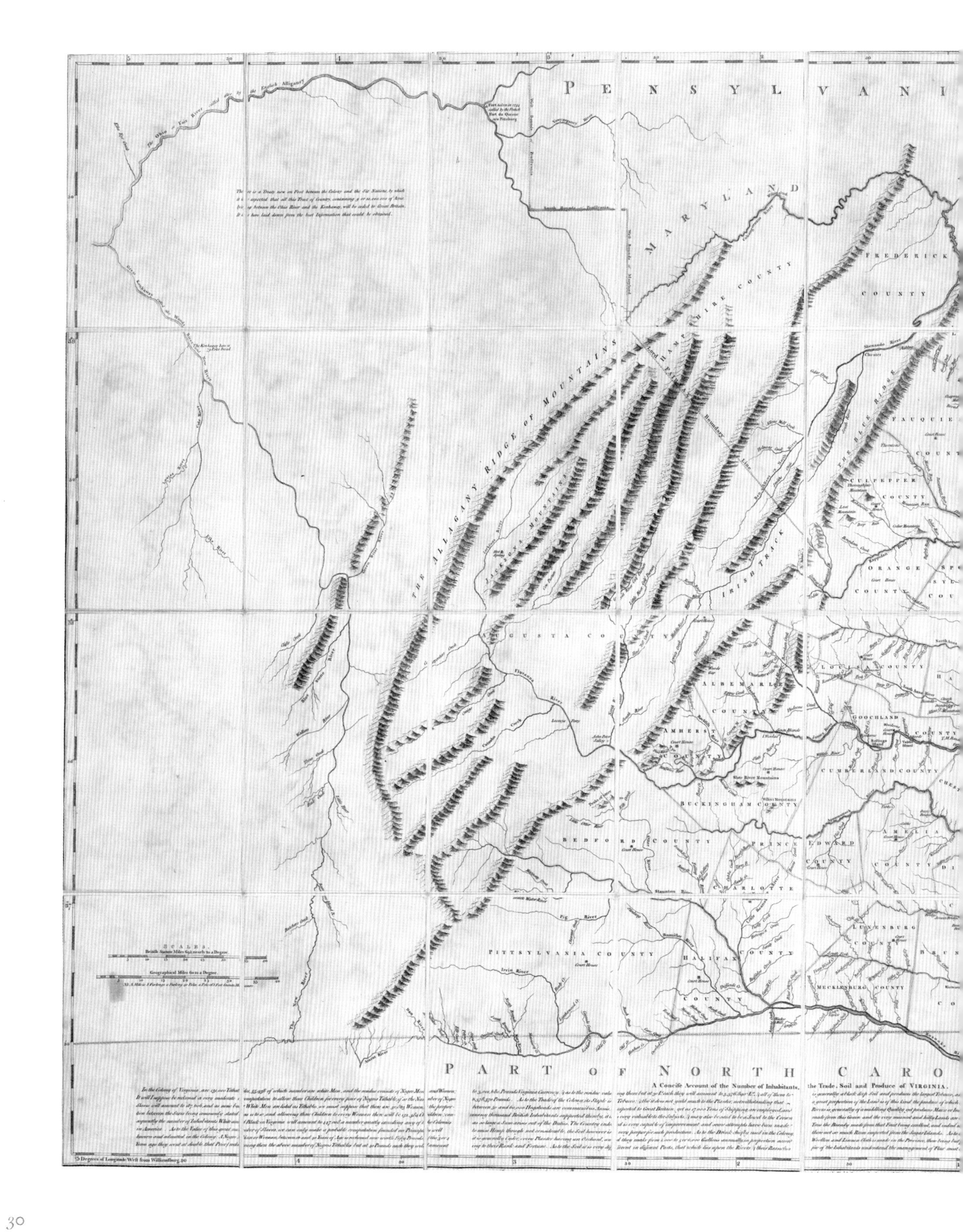

PENSYLVANI
MARYLAND
FREDERICK
COUNTY
HAMPSHIRE COUNTY
THE ALLAGANY RIDGE OF MOUNTAINS
FAUQUIER
CULPEPPER
COUNTY
ORANGE
COUNTY
IRISH TRACK
AUGUSTA COUNTY
ALBEMARLE
COUNTY
LOUISA COUNTY
GOOCHLAND
AMHERST
COUNTY
CUMBERLAND COUNTY
BUCKINGHAM COUNTY
BEDFORD COUNTY
PRINCE EDWARD
COUNTY
AMELIA
CHARLOTTE
LUNENBURG
MECKLENBURG COUNTY
PITTSYLVANIA COUNTY
HALIFAX COUNTY
SCALES
British Statute Miles 69½ nearly to a Degree
Geographical Miles 60 to a Degree
PART OF NORTH CARO
A Concise Account of the Number of Inhabitants, the Trade, Soil and Produce of VIRGINIA.

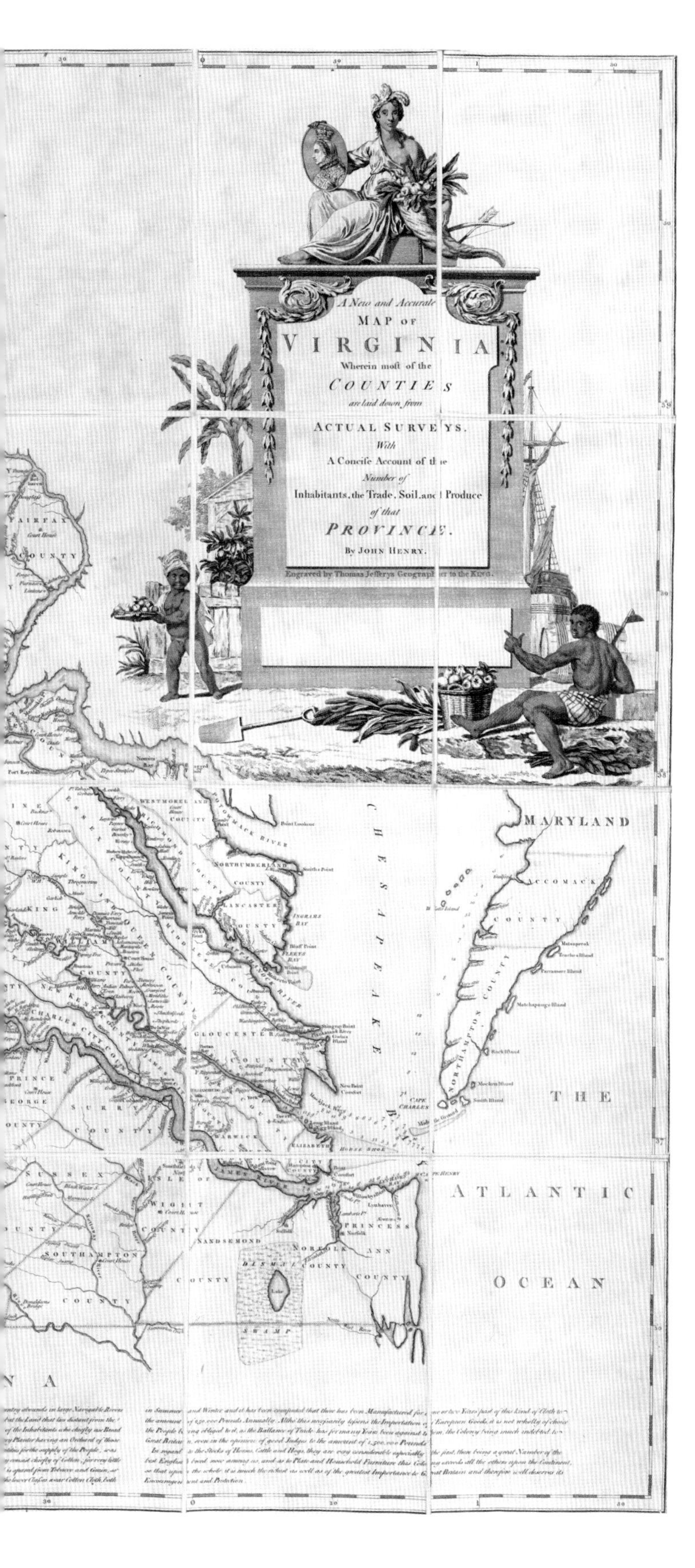

A new and accurate map of Virginia wherein most of the counties are laid down from actual surveys. With a concise account of the number of inhabitants, the trade, soil, and produce of that Province.

This 1770 map by John Henry indicates the Fairfax boundary line and names the Blue Ridge Mountains. Each county has a courthouse, and ordinaries (inns) have been built for travelers. Settlers have moved west of the Proclamation Line of 1763, past which George III forbade westward settlement to avoid conflict with the American Indians.

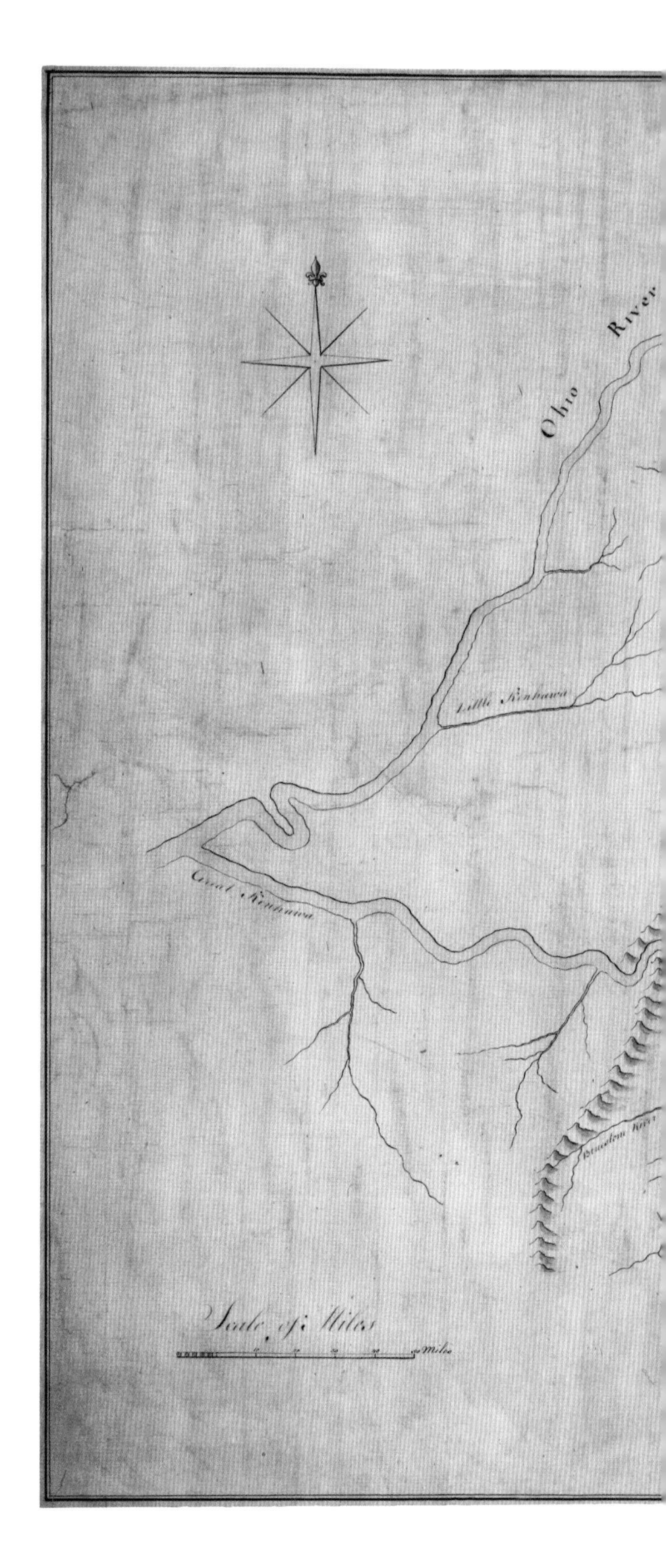

A map of Potomack and James rivers in North America shewing their several communications with the navigable waters of the new province on the river Ohio.

This map by John Ballendine, ca. 1773, renders the Potomac and James Rivers and their relationship to the "river Ohio," which is indicated by a wavy line at left. This "communication" would be by wagon road, as canals had not yet been built. Fort Pitt, which had been the French Fort Duquesne; Fort Cumberland; and Fort Pleasant are identified.

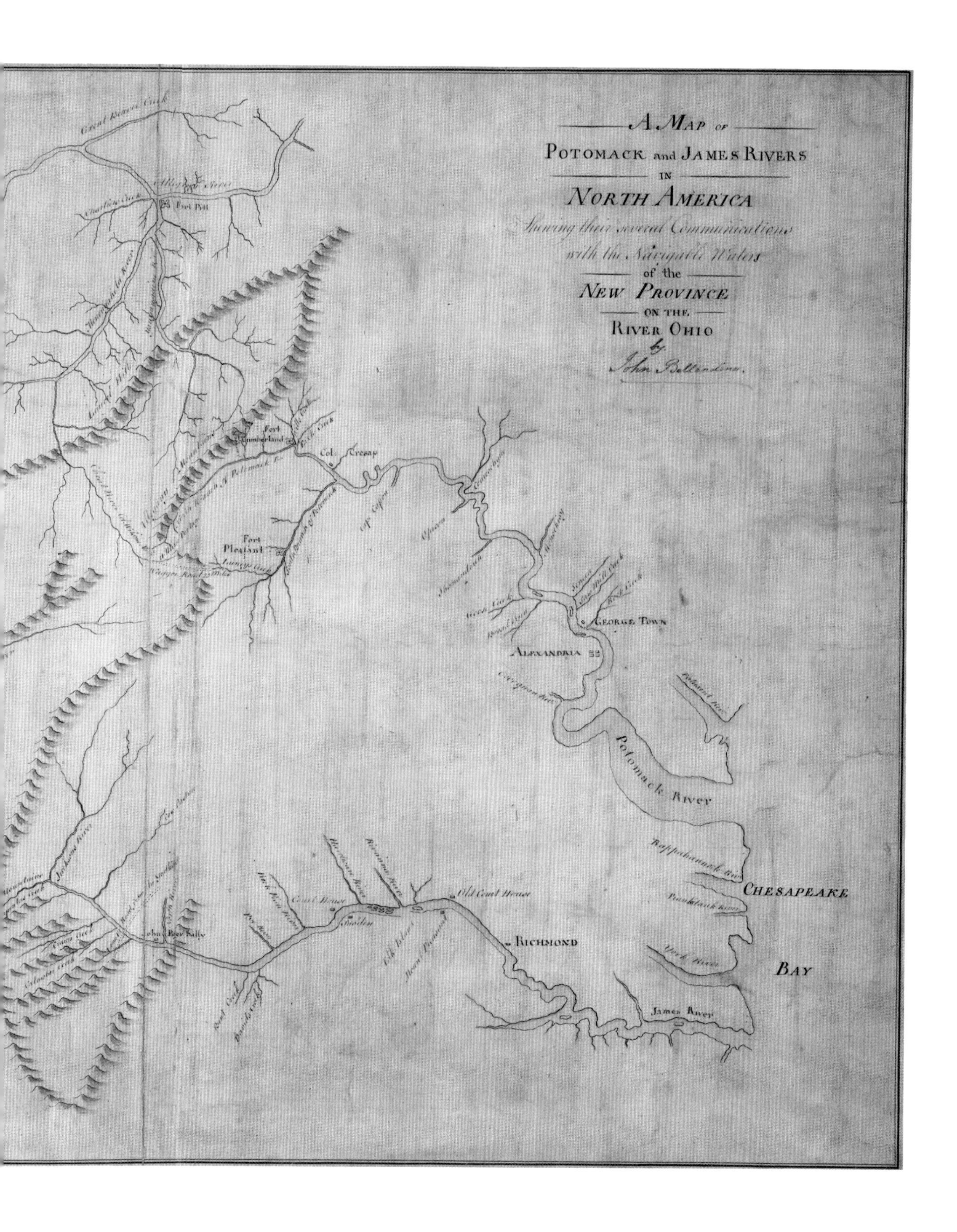
A MAP OF
POTOMACK and JAMES RIVERS
IN
NORTH AMERICA
Shewing their several Communications
with the Navigable Waters
of the
NEW PROVINCE
ON THE
RIVER OHIO
by
John Ballendine.
Fort Cumberland
Col. Cresap
Fort Pleasant
GEORGE TOWN
ALEXANDRIA
Potomack River
Rappahanock River
CHESAPEAKE
BAY
York River
James River
RICHMOND
Old Court House
Court House

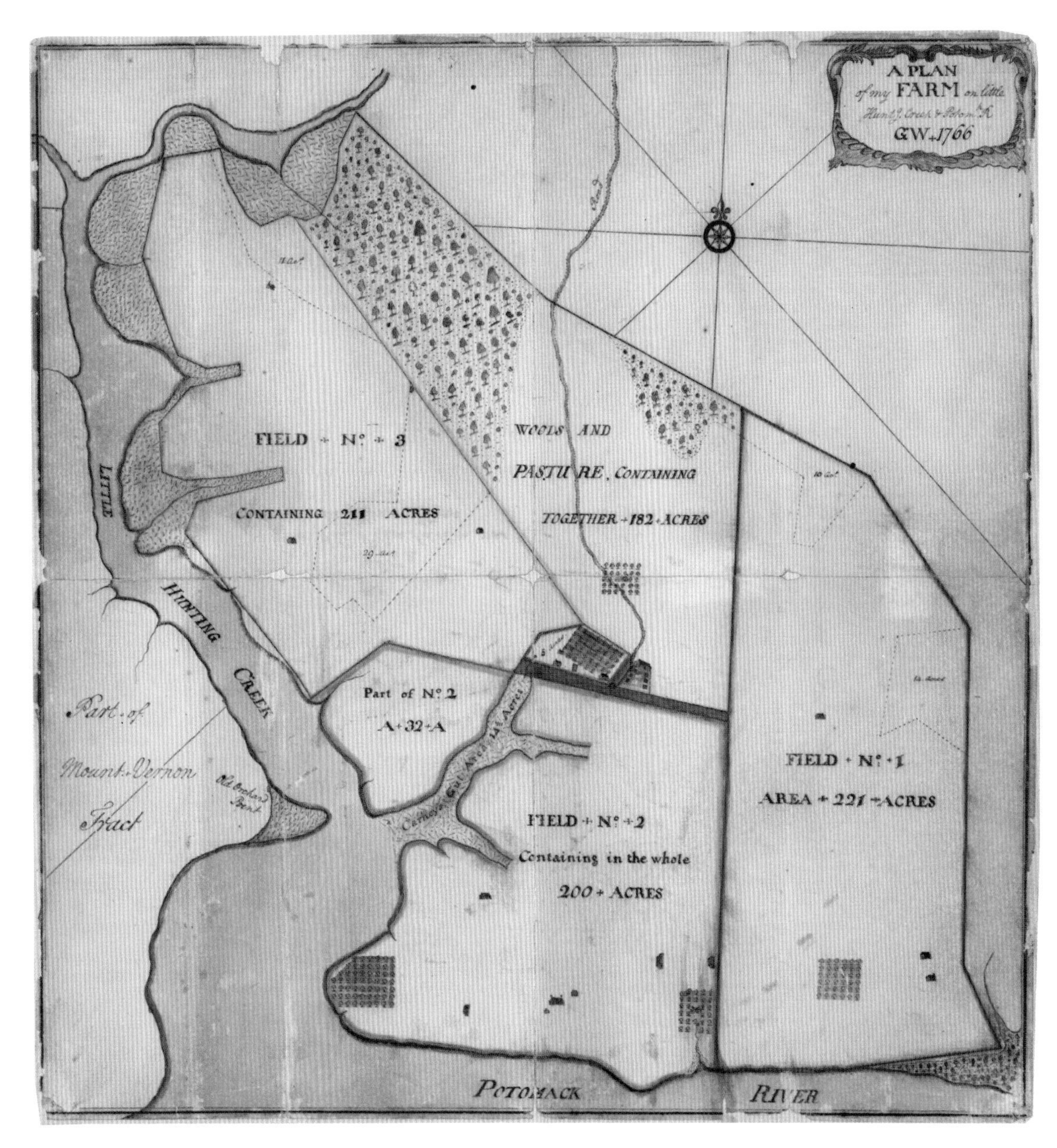

A plan of my farm on Little Huntg. Creek & Potomk. R. G. W. George Washington's farm on Little Hunting Creek and the Potomac River, drawn by Washington himself in 1766.
At this time, after the French and Indian War and before the American Revolution, Washington was a successful planter. The various fields are similar in size, just over 200 acres, and Washington has left 182 acres in woodland and pasture, an indication of his care for his land.

The Revolutionary War and the Expansion of the United States

The causes that led Virginia and her sister colonies to declare independence went back almost to the first settlement. Separated by an ocean from England, they were alternately ignored, rewarded, punished—and almost always mismanaged.

They were established to make England richer. When no gold was found, they were to provide raw materials—timber, furs, iron ore, and pitch—and to buy manufactured goods from English merchants. Many royal governors never came to America but paid a substitute. Colonial petitions to the king often took years to be seen by the monarch, only to be rejected.

Virginians began to think of themselves as Americans, not Englishmen. They wanted to govern themselves, to trade with other nations, to feel safe in their homes, and to have some choice as to who landed on their shores.

Military problems began with the French when they moved into what is now western Pennsylvania, then Virginia. Young George Washington was sent with militia on a failed mission to scare off the French, and next with Gen. Edward Braddock's British troops to take the French fort, Duquesne. Through Braddock's arrogance and inexperience in the wilderness, he and his army were ambushed and slaughtered. Washington survived to make a successful trip to the area with British general John Forbes. The French evacuated their fort and left the area, but the British-French conflict spread worldwide. Britain won, and in the peace treaty of 1763, France withdrew from her territory in America, except for Louisiana.

In the interim between 1758 and 1763, American Indians attacked freely in Virginia's western territories, and settlers fled back across the mountains. Virginia built a chain of forts in the Shenandoah Valley, with Washington in charge. King George's solution was to forbid further settlement beyond the mountains.

To pay off its huge war debt, Britain strengthened the Navigation Acts and added a sugar tax and a stamp tax, the latter requiring a British-issued stamp on every legal document, including marriage licenses, college diplomas, and bills of lading for ships. In short, nearly everyone would be taxed. Virginians boycotted and threatened the tax collectors. The Stamp Act was repealed, but Parliament declared that it had the right to tax the colonies and soon passed the Townsend Acts, taxing lead, glass, paint, and tea. Again, these taxes applied to almost everyone.

After Boston destroyed shiploads of tea in 1773 and the British closed their port, Virginia sent food to the beleaguered Bostonians and called for a day of prayer. John Murray, the Earl of Dunmore, Virginia's governor, forbade it and dissolved the House of Burgesses, the colony's legislature. He had already fomented war with the Native Americans on the frontier through his land agent. Virginians joined other colonists in the First Continental Congress in Philadelphia, which declared unity in banning all British goods. They stopped short of organizing an army.

When Virginians met in March 1775 to elect delegates to the Second Continental Congress, Patrick Henry introduced a resolution that Virginia should arm itself for war. The delegates, while stunned, voted yes. A month later, war began—in Massachusetts. Governor Dunmore had the Royal Marines from nearby Norfolk come secretly at night and take away the gunpowder from the colony's Powder Horn. He barricaded himself in the Governor's Mansion but eventually called the Burgesses back into session, telling them that Parliament had repealed all taxes except those to regulate trade, on the condition that the colony tax itself for the defense of the empire. It was too late for concessions, however—Virginians now wanted independence.

Sensing their mood, Dunmore and his family fled to a ship anchored at Norfolk. He gathered an army, raided nearby farms, and freed all indentures and slaves, offering arms to any who would join him. After his army was defeated in the Battle of Great Bridge on December 9, 1775, he withdrew to a ship, and on New Year's Day 1776, he bombarded Norfolk, leveling the city and burning its warehouses, before sailing away.

Without a governor, Virginia was free to govern itself. Delegates meeting in Williamsburg in June 1776 accepted George Mason's Declaration of Rights and made it a part of their new constitution. Patrick Henry became Virginia's first post-colonial governor.

Meanwhile, in Philadelphia, George Washington had been given command of the American army, and the Declaration of Independence was written and signed.

For the first years of the Revolutionary War, Virginians mostly saw action in the Ohio Territory. Then, in 1779 Virginia was invaded by British troops. Tobacco and supplies were destroyed at Portsmouth, and Suffolk was burned. The state had few defenders, as forty thousand Virginia troops were serving on other battlefields. Governor Thomas Jefferson moved the capital to Richmond.

The war had dragged on for five heroic years, in New York, New England, and the Carolinas, as Washington struggled with lack of supplies, lack of money, and a surplus of illnesses and desertions among his troops. There was little military action in Virginia. Then, in 1780 Benedict Arnold, who had deserted to join the British, sailed up the James River, destroyed supplies and buildings around Richmond, and returned to Portsmouth for his winter encampment, looting as he went.

In early 1781 Gen. Charles Cornwallis moved his army from the Carolinas to the Virginia peninsula at Yorktown and sent his cavalry commander, Banastre Tarleton, known as "the butcher," to seize Jefferson, who was staying at Monticello. Warned by John Jouett, Jefferson escaped minutes before the British army arrived.

Washington and French general Rochambeau marched their forces south from New York to Virginia to join the army of French nobleman Marquis de Lafayette, and the French fleet sailed from the West Indies. The British fleet, which was bringing supplies and reinforcements to General Cornwallis, could not enter Virginia waters. With escape cut off, Cornwallis had no choice but to surrender, which he did on October 19, 1781. The largest army in North America laid down its arms before a bedraggled but valiant group of Americans.

It was obvious that this was the end of the war, though New York, Charleston, and Savannah were still under British control, and some isolated battles took place over the next year. In 1783 a peace treaty was signed recognizing the independence of the thirteen states, though British troops held some forts in Ohio, which was to lead to another war.

Some Virginians had supported the British king, and when the break came, they faced hard choices that split friendships and even families. Loyalist John Saunders joined Dunmore at Norfolk and fought with British troops. After the war he joined other Loyalists in Canada and became one of the founders of New Brunswick. Washington's friend George William Fairfax went to England, as did John Randolph, a member of the Burgesses. His son Edmund stayed in Williamsburg and became America's first attorney general and later secretary of state.

Life was chaotic in the former colonies. They were independent but broke and indebted. Estates had been ruined, soldiers hadn't been paid, and the states argued over boundaries. Maryland had earlier refused to sign the Articles of Confederation, a loose union of the states, until Virginia relinquished her Northwest Territory. Maryland, a landlocked state, said that it had no land to pay its soldiers with and therefore might have to recall them. Virginia suspected that land speculators were bribing members of Congress and bringing spurious claims of ownership although the land had been granted to Virginia in 1609. In 1784 Virginia unselfishly gave a quarter million square miles of land to Congress by Deed of Cession, requiring that it be for the benefit of the nation as a whole and that it be made into states equal in status to the original thirteen.

Problems continued. Each state issued its own money and threw up trade barriers against each other. Realizing the need for unity, delegates met first at Mount Vernon, then at Annapolis, and finally at Philadelphia, where they wrote the Constitution and established a nation. Washington became the first president, Jefferson the first secretary of state. A new, permanent capital was to be built on the Potomac, on Virginia's border. In return for this, Virginia agreed to assume a share of other states' debts, though she had paid much of her own.

Virginians generally were in favor of becoming part of a nation. As the largest state, Virginia would have power for the next three decades, being the home state of three of the first four presidents. But as states were formed from her western territory, Virginia's influence lessened. Primarily rural and agricultural, Virginia depended on slave labor and the salability of her crops, but planters still needed to buy manufactured goods, made primarily in the North and shipped on northern vessels. New immigrants were more likely to settle in a northern city than in the South, for the cheap land that had lured earlier settlers to the South was gone.

By 1820, New York had a population of 1,372,812, surpassing Virginia's 1,065,336, which included nonvoting slaves. The scepter of power had passed.

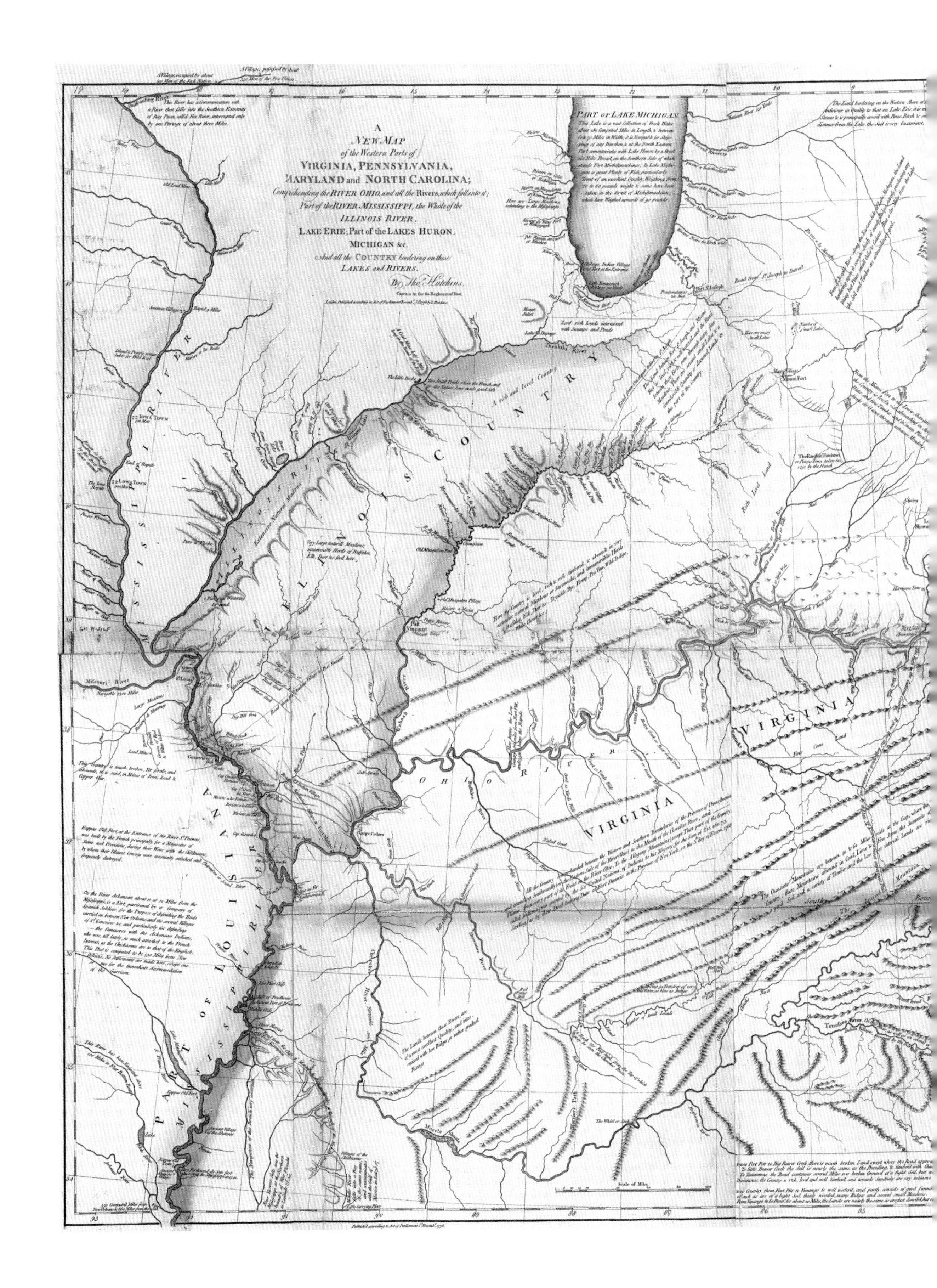
A
NEW MAP
of the Western Parts of
VIRGINIA, PENNSYLVANIA,
MARYLAND and NORTH CAROLINA;
Comprehending the RIVER OHIO, and all the Rivers, which fall into it;
Part of the RIVER MISSISSIPPI, the Whole of the
ILLINOIS RIVER,
LAKE ERIE; Part of the LAKES HURON,
MICHIGAN &c.
And all the COUNTRY bordering on these
LAKES and RIVERS.
By Thos. Hutchins,
Captain in the 60 Regiment of Foot.
PART OF LAKE MICHIGAN
ILLINOIS COUNTRY
ILLINOIS RIVER
MISSISSIPPI RIVER
PART OF LOUISIANA
OHIO RIVER
VIRGINIA
Missouri River
Scale of Miles

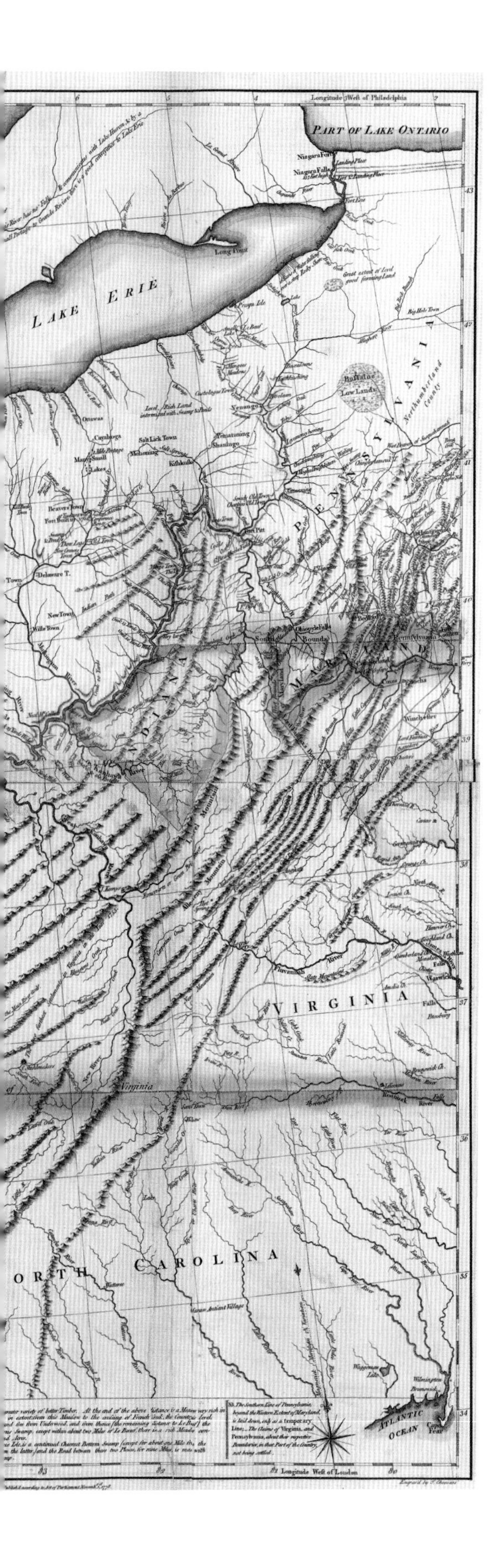

A new map of the western parts of Virginia, Pennsylvania, Maryland, and North Carolina; comprehending the River Ohio, and all the rivers, which fall into it; part of the River Mississippi, the whole of the Illinois River, Lake Erie; part of the Lakes Huron, Michigan &c. and all the country bordering on these lakes and rivers, by Thos. Hutchins, Engrav'd by T. Cheevers.

Published in 1778, during the Revolutionary War, this Thomas Hutchins map could have been useful to both American and British troops, as it shows the Ohio and its tributaries, the Illinois, part of the Mississippi, and the Great Lakes in detail. A notation indicates the Ohio Valley ". . . is level and rich, abounding in streams of water and fine timber and in many places the road directs its course through extensive meadows." Of present Indiana, Hutchins noted there were "great herds of buffalo, elk and deer," and in what became West Virginia and western Pennsylvania, he located "coals."

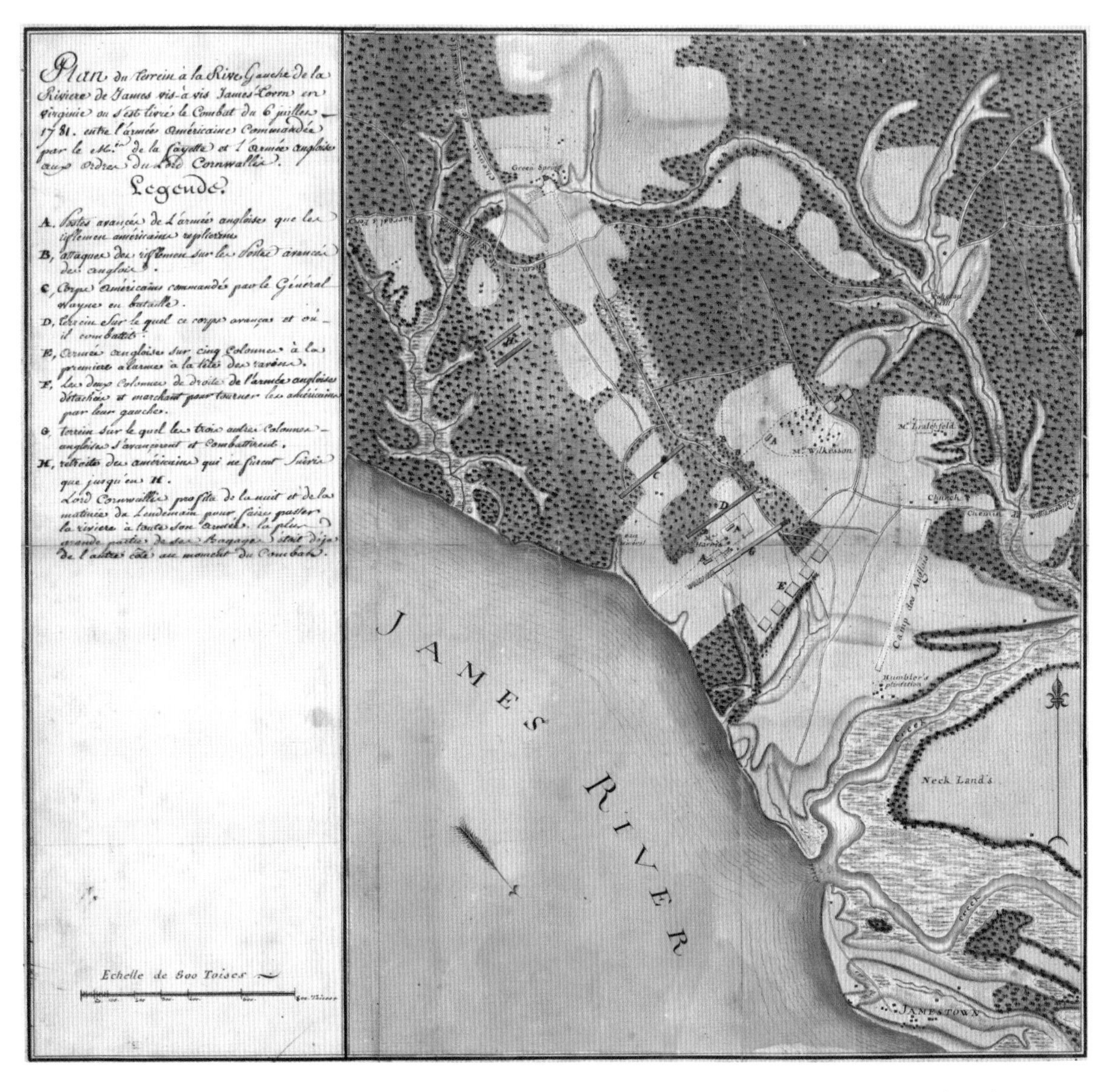

Plan du terrein à la rive gauche de la rivière de James vis-à-vis Jamestown en Virginie ou s'est livré le combat du 6 juillet 1781 entre l'armée américaine commandée par le Mis. de La Fayette et l'armée angloise aux ordres du Lord Cornwallis. [Signé:] Desandroüins. Desandroüins, Jean Nicolas, 1729–1792.

A map drawn in July 1781 by Jean Nicolas Desandroüins showing the relative positions of the British, French, and American troops along the James River near Jamestown. Cornwallis arrived in the area with his army in early July 1781 and remained until his surrender in October. *A,* in the center, is the position of the British troops; the French were opposite the British, separated by a wooded area, at *D;* the Americans were at *B* and *C.* The British were boxed in, prevented from retreating by water or land.

Map of the country between Albemarle Sound, and Lake Erie, comprehending the whole of Virginia, Maryland, Delaware and Pensylvania, with parts of several other of the United States of America. Engraved by S. J. Neele.

In this 1787 map, Thomas Jefferson used the Hutchins map (see page 39) of the Ohio Valley maps drawn by his father and Joshua Fry for the area east of the Allegheny Mountains, as well as Scull's map of Pennsylvania, with additions " . . . where they could be made on sure ground." Virginia had ceded her western territories to the new government of the United States, and Jefferson wanted to show the extent of the ceded territory, which could be made into new states.

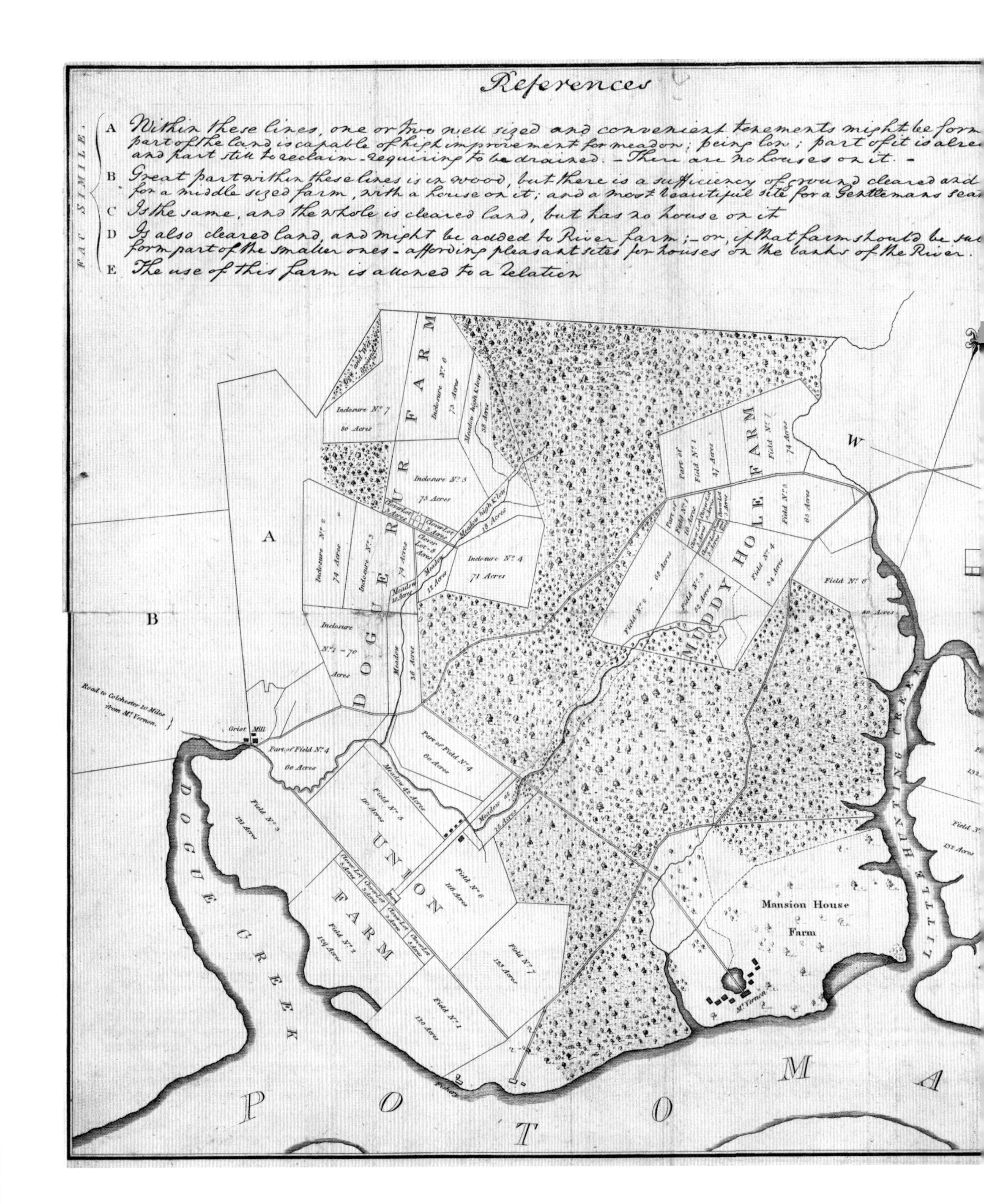

References
FAC SIMILE.
A Within these lines, one or two well sized and convenient tenements might be form
part of the land is capable of high improvement for meadow; being low; part of it is alre
and part still to reclaim - requiring to be drained. - There are no houses on it. -
B Great part within these lines is in wood, but there is a sufficiency of ground cleared and
for a middle sized farm, with a house on it; and a most beautiful site for a Gentlemans sea
C Is the same, and the whole is cleared land, but has no house on it
D Is also cleared land, and might be added to River farm; - or, if that farm should be sa
form part of the smaller ones - affording pleasant sites for houses on the banks of the River.
E The use of this farm is allowed to a Relation
DOGUE RUN FARM
MUDDY HOLE FARM
UNION FARM
Mansion House Farm
Mt. Vernon
Grist Mill
Road to Colchester 10 Miles from Mt. Vernon
DOGUE CREEK
LITTLE HUNTING CREEK
POTOMA
Fishery

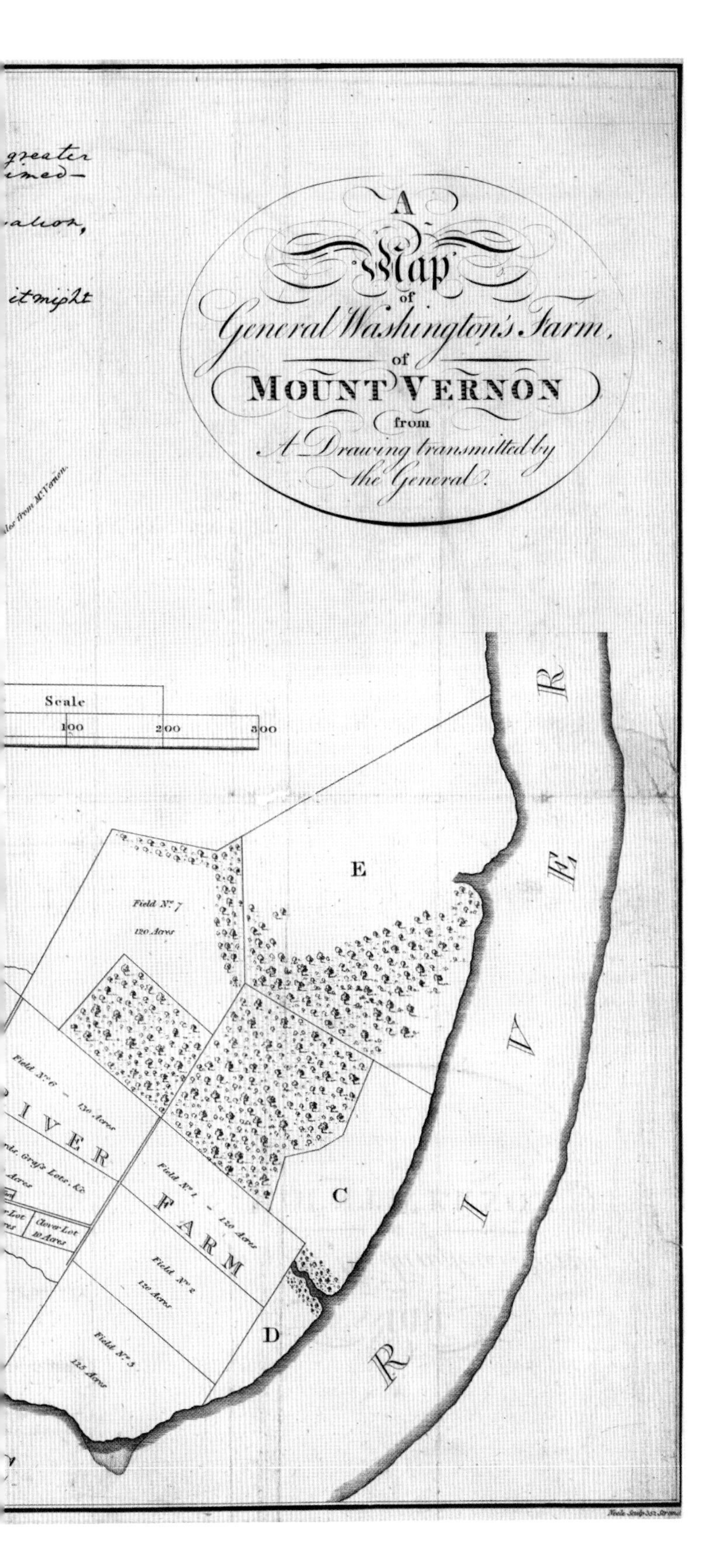

A map of General Washington's farm of Mount Vernon from a drawing transmitted by the General.

George Washington drew this map of his Mount Vernon farm soon after retiring from the presidency in 1796. He wrote of *A*: "Within these lines one or 2 well-sized and convenient tenements might be formed." A portion, he wrote, had been reclaimed, and part of it needed draining. Of *B* (far left) he wrote that there is no house on it, as it is mostly in woodland, but there is "a most beautiful site for a gentleman's seat."

View of the University of Virginia, Charlottesville & Monticello, taken from Lewis Mountain, drawn from nature & print. in colors by E. Sachse & Co.

In this view of the University of Virginia, Charlottesville, and Monticello, drawn by Casimir Bohn in 1856, Monticello is on the far hilltop at top right center. The Rotunda is shown in side view, with the classical ranges and pavilions in rows before it. This publicly funded university, designed by Thomas Jefferson, was the first secular university in America. It opened in 1825, a year after Jefferson's death.

The Antebellum South

As early as 1699, Virginia had tried to slow the slave trade by imposing a heavy tax on the importation of slaves, but the English government disallowed such a tax as being unfair to British ship owners. So the trade continued. The U.S. Constitution provided for the end of the slave trade in 1808, thus ending the dreadful "triangular trade" in which two-thirds of the Africans perished on the voyage to America. By 1810, one in ten blacks in Virginia was free. Some blacks worked to earn their freedom, and others were freed per their owners' wills.

In 1776 Thomas Jefferson proposed freeing Virginia's slaves. But who would pay for the planters' loss, and what would become of the freed blacks? Following the Revolutionary War, slaves were given more opportunities to worship, to gather socially, and to be educated. Then word came that blacks had overthrown their masters in Haiti, and Virginia slaves began to wonder if they could do likewise. Slave owners, who lived in fear of such an uprising, wondered, too.

The answer came in August 1800. Gabriel Prosser, a slave in Henrico County, planned a statewide insurrection. He and his followers would kill families of slave owners, kidnap Governor James Monroe, and open the penitentiaries. The plot was discovered, the militia called out, and a storm postponed the uprising. The leaders were found and executed.

Worried Virginians looked for solutions. One was the American Colonization Society, which purchased slaves and sent them to West Africa, where they founded the nation of Liberia. Most blacks had no wish to leave, and the cost of buying and settling them was high. Virginia had one notable success at this endeavor, however: Joseph Jenkins Roberts, from Petersburg, emigrated to Liberia with his mother and siblings. From there, he and an educated black clergyman from Richmond, William Colson, exported African products to New York and Philadelphia, and they sold American-made goods in Liberia. When Liberia became a nation, Roberts was elected its first president.

Virginian John Randolph predicted that slavery would gradually die out as uneconomic, if the North did nothing. He pointed out that slavery had existed in the northern states but didn't any longer, and it was about finished in Maryland.

Virginia would be the next. In 1831 the Virginia General Assembly proposed using tax funds to buy the state's slaves and free them. The measure came close to passage, despite its cost.

Then, that same year, Nat Turner, an educated slave, organized an insurrection in Southampton County that killed fifty-eight whites. Thereafter it was illegal to teach slaves to read and write, and whenever they left their owner's plantation, they had to carry a pass. During the 1830s, several years of low crop prices made slaves less economical, and in that decade alone, 118,000 Virginia slaves were sold to cotton growers in the Deep South.

Virginia was separating into two societies: the tidewater planters who depended mainly on tobacco, and the western settlements of small farmers and merchants. Power still lay with the tidewater group, who shipped their crops through Norfolk and saw no need to be taxed for internal improvements that would benefit the west. Most taxes were tariffs on imported goods. Such tariffs, which reached 33 percent in 1824, paid for roads and canals to the west and protected northern manufacturing from overseas competition.

Maryland, on the other hand, saw the economic advantage of connecting the new states of the Ohio Territory—Ohio, Indiana, and Illinois—to the Atlantic ports and joined in building the Chesapeake and Ohio Canal. Virginia missed the opportunity to become a shipping power and instead built the James River and Kanawha Canal, joining Richmond to the west. It was proposed to extend into what is now West Virginia and to be joined by wagon road to Point Pleasant on the Ohio River, but it was only completed as far as Buchanan. Later, when railroads were built, the Chesapeake and Ohio followed Maryland's route, from Cumberland, Maryland, to Vandalia, Illinois.

Virginia planters were unable to invest in transportation or industry, for their wealth was tied up in land and slaves. In the pre–Civil War period, Virginia land was selling for $150 an acre, and the average price of a slave was $1,000. There was some industry, such as the Tredegar Iron Works in Richmond; flour, iron, and cotton mills in Petersburg, which also became a railroad terminus; and cotton mills in Danville. Still, there was little financial opportunity for the ambitious young, and some of Virginia's ablest moved away. Four examples are Sam Houston, who became governor of Tennessee and then president of the Republic of Texas; Stephen F. Austin, who helped found the state of Texas; William Claiborne, who became governor of Louisiana; and Thomson Mason, who at nineteen was appointed governor of Michigan Territory. Over three hundred thousand Virginians left the state during the antebellum period, and few immigrants were arriving. How could paid labor compete with slave labor?

In 1834 a young Virginia inventor, Cyrus McCormick, sold his first reaper. This invention forever changed the way wheat was harvested, making inexpensive bread available worldwide and thus allowing farmers to instead work in manufacturing jobs. For a time, Virginia was one of the top four wheat-growing states, and McCormick became a millionaire—but not in Virginia. Because of lengthy transportation—by wagon to the James River, out through the port of Norfolk, and upriver from New Orleans—to reach the plains of the Midwest, McCormick moved his manufacturing to Chicago.

Virginia was stuck with slavery and began to rationalize it: Slaves were cared for whether they were young or old, sick or well, whereas paid workers were not. Slaves paid no rent yet could always be sure of shelter.

As states were added to the nation, Virginia's representatives in Congress foresaw that they would soon be outvoted. Their tax money would be used to help the North and the West, and their way of life would be destroyed. The solution was to keep the number of free and slave-owning states equal so that, at least in the U.S. Senate, they could protect their interests. The Missouri Compromise of 1820, admitting Missouri as a slave state and Maine as a free state, put the matter to rest for twenty years.

Literature flourished in antebellum Virginia. The *Southern Literary Messenger* was read all over the United States, not just in Virginia, and introduced Edgar Allan Poe's work to the world. But Virginia had few large cities to contribute to the development of a broad literary culture. It was still agricultural, and for a time agriculture improved and flourished, thanks to Thomas Jefferson's invention of a better plow and Edmund Ruffin's demonstrations on the benefits of fertilizer. But the fortunes and the prestige of the Old Dominion fell. Jefferson sold his library to the Library of Congress and died in poverty. Washington's home, Mount Vernon, was derelict when a group of women purchased it for $200,000 and opened it as a national shrine. Once-flourishing Alexandria, home of Washington, Fairfax, and the Lees, was almost a ghost town.

Virginia had a brief moment of national glory again in 1840, when both candidates for the presidency and vice presidency, William Henry Harrison and John Tyler, were Virginia-born. When Harrison died after only a month in office, Tyler became president. He hoped to bring North and South to the center politically. He vetoed tariffs and the national bank bill, and he planned to annex Texas as a slave-holding state. Tyler was considered a traitor to the Whig party and retired after one term.

Pushed by its residents in western Virginia, the General Assembly wrote a new constitution in 1850, giving the vote to all white males regardless of property ownership and providing for direct election of the governor, who had been appointed by the legislature. The districts were reapportioned, giving more representation to western residents, and in 1852 the first Virginia governor from west of the mountains, Joseph Johnson, was elected.

Jefferson had described slavery as a "fire bell that rings in the night." It rang for the South in 1859 when a fanatic abolitionist, John Brown, gathered weapons and followers and attacked the arsenal at Harpers Ferry, Virginia. His control was brief. The marines, the Virginia militia, and the U.S. Army under Robert E. Lee routed him, but not before he had taken hostages and killed several people, including the mayor of Harpers Ferry. He planned a slave uprising for the entire country, using the weapons he'd seize from the arsenal. Brown was tried, found guilty, and hanged.

Feelings hardened between North and South when Brown became a hero in northern newspapers and it was revealed that religious groups in the North had raised money to support his attack.

Compromise was less and less likely.

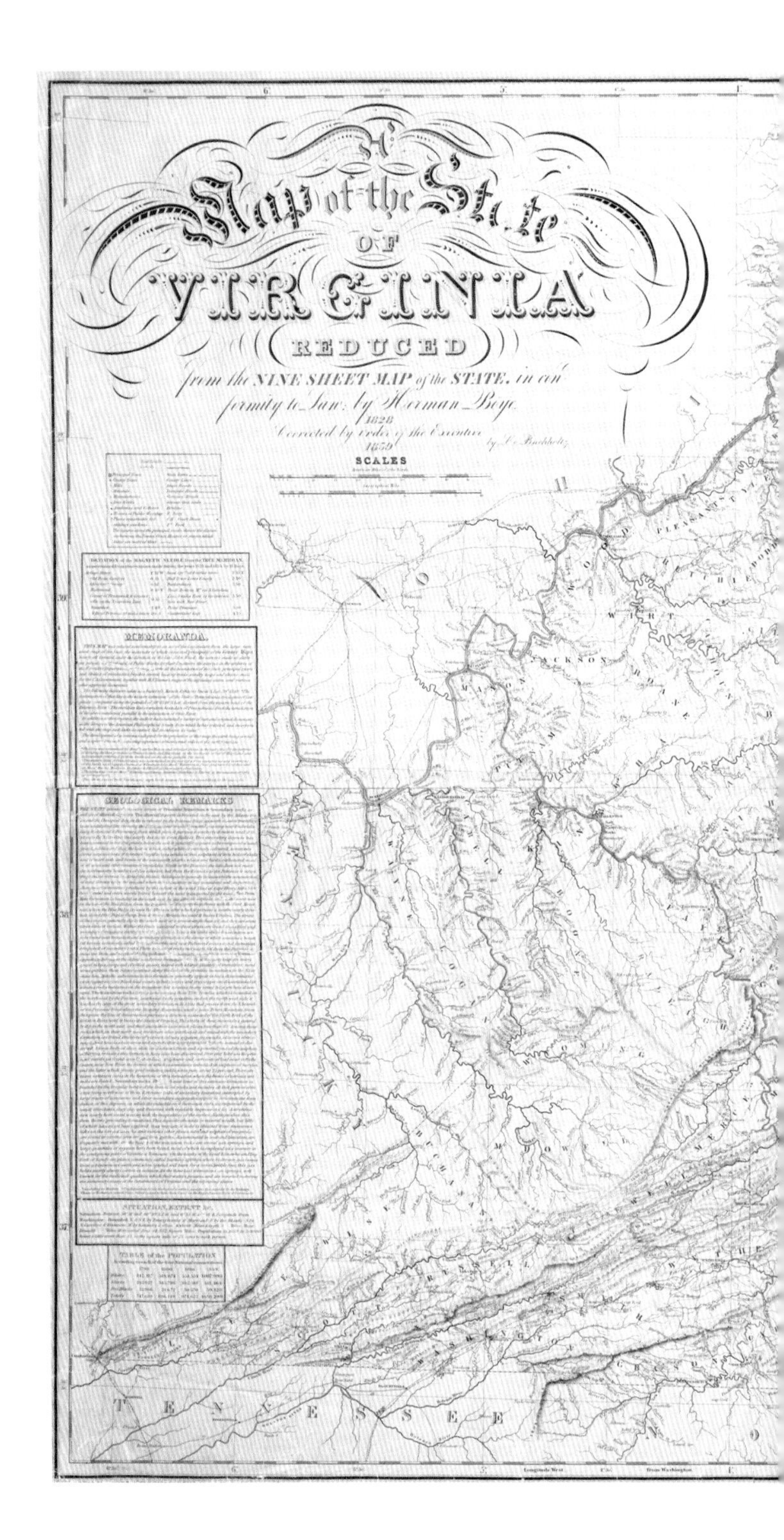

A map of the state of Virginia: reduced from the nine sheet map of the state in conformity to law, by Herman Böÿe, 1828.

Herman Böÿe assembled this map from the nine-sheet map of Virginia required by law in 1827. Attention and settlement have moved west. Jamestown is not indicated, but all the counties of Virginia, including those soon to become West Virginia, are labeled. The counties of Kentucky and Tennessee have become states. The map is detailed, showing population and the steamboat routes on the major rivers.

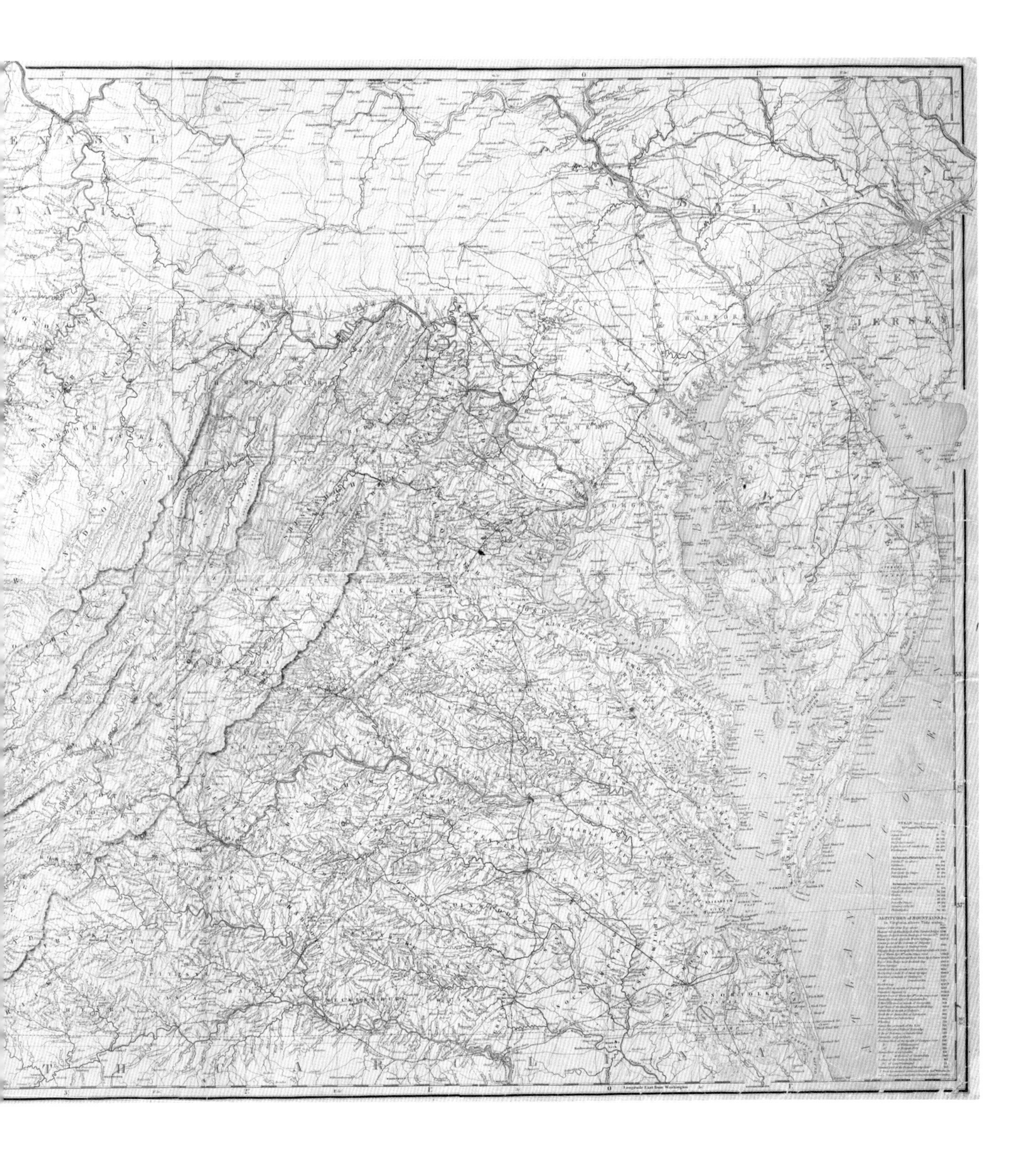

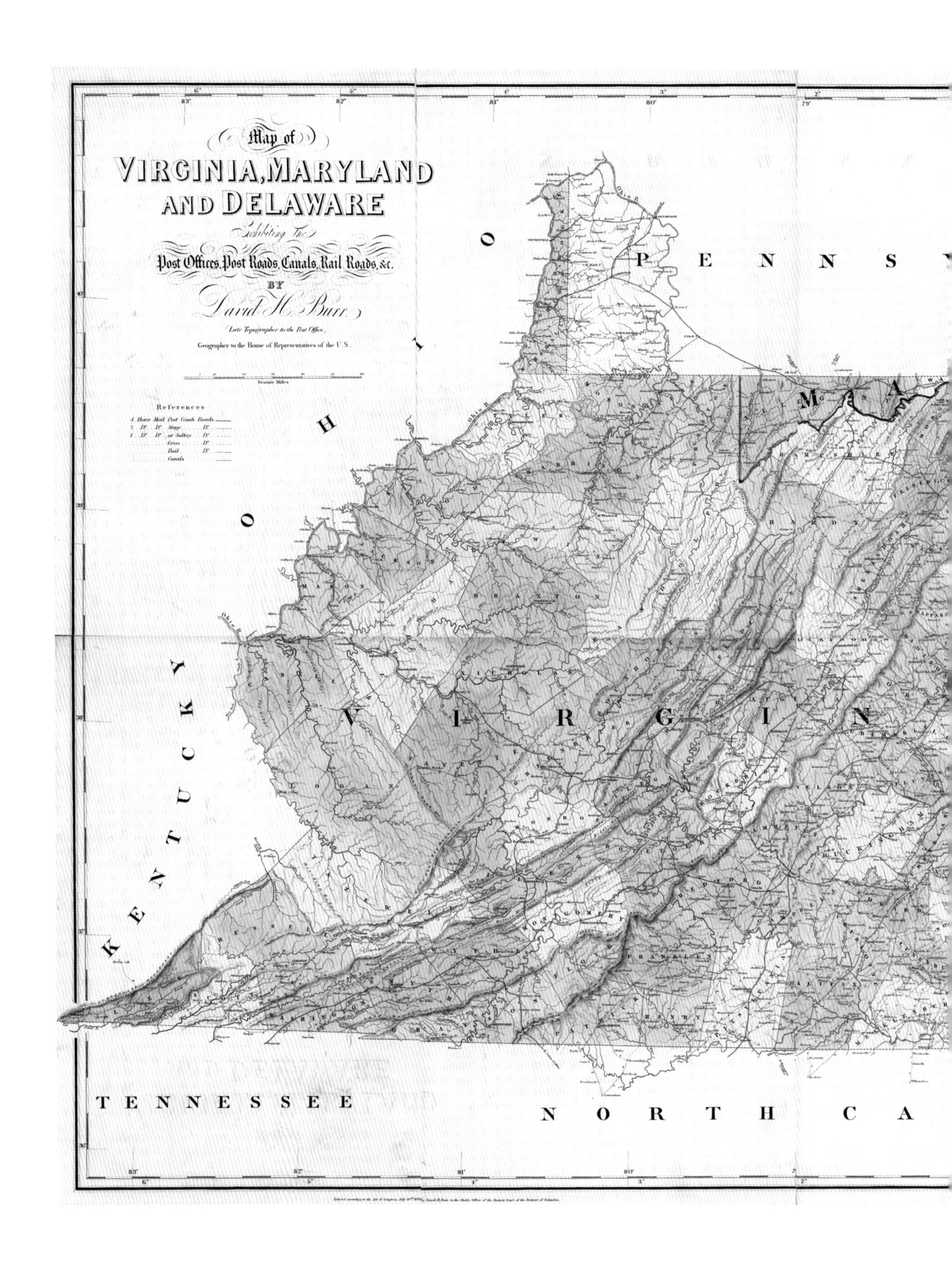
Map of
VIRGINIA, MARYLAND
AND DELAWARE
Exhibiting The
Post Offices, Post Roads, Canals, Rail Roads, &c.
BY
David H. Burr.
Late Topographer to the Post Office.
Geographer to the House of Representatives of the U.S.
Statute Miles
References
4 Horse Mail Post Coach Roads
2 Do. Do. Stage Do.
1 Do. Do. or Sulkey Do.
Cross Do.
Rail Do.
Canals
O H I O
P E N N S Y
K E N T U C K Y
V I R G I N
T E N N E S S E E
N O R T H C A

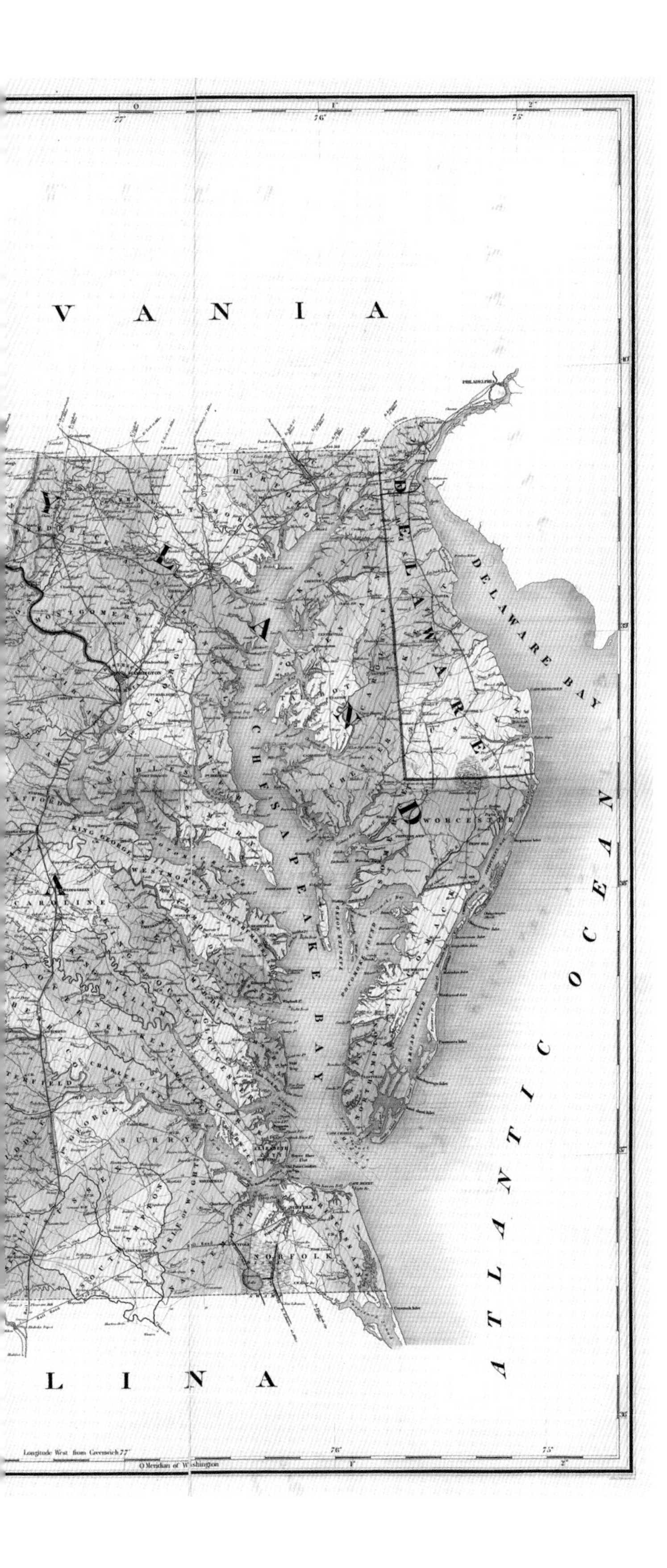

Map of Virginia, Maryland and Delaware exhibiting the post offices, post roads, canals, rail roads &c. by David H. Burr (late topographer to the Post Office), geographer to the House of Representatives of the U.S.

Washington, D.C., is shown as a square on both sides of the Potomac River; Arlington was returned to Virginia in the 1840s. Post roads, such as that from D.C. to Fredericksburg and Richmond, became the basis of I-95, while another, from Richmond to Charlottesville and Staunton, became the route of I-64. County seats were labeled, including the ironically named Union County in what soon became West Virginia when it joined the union in 1863.

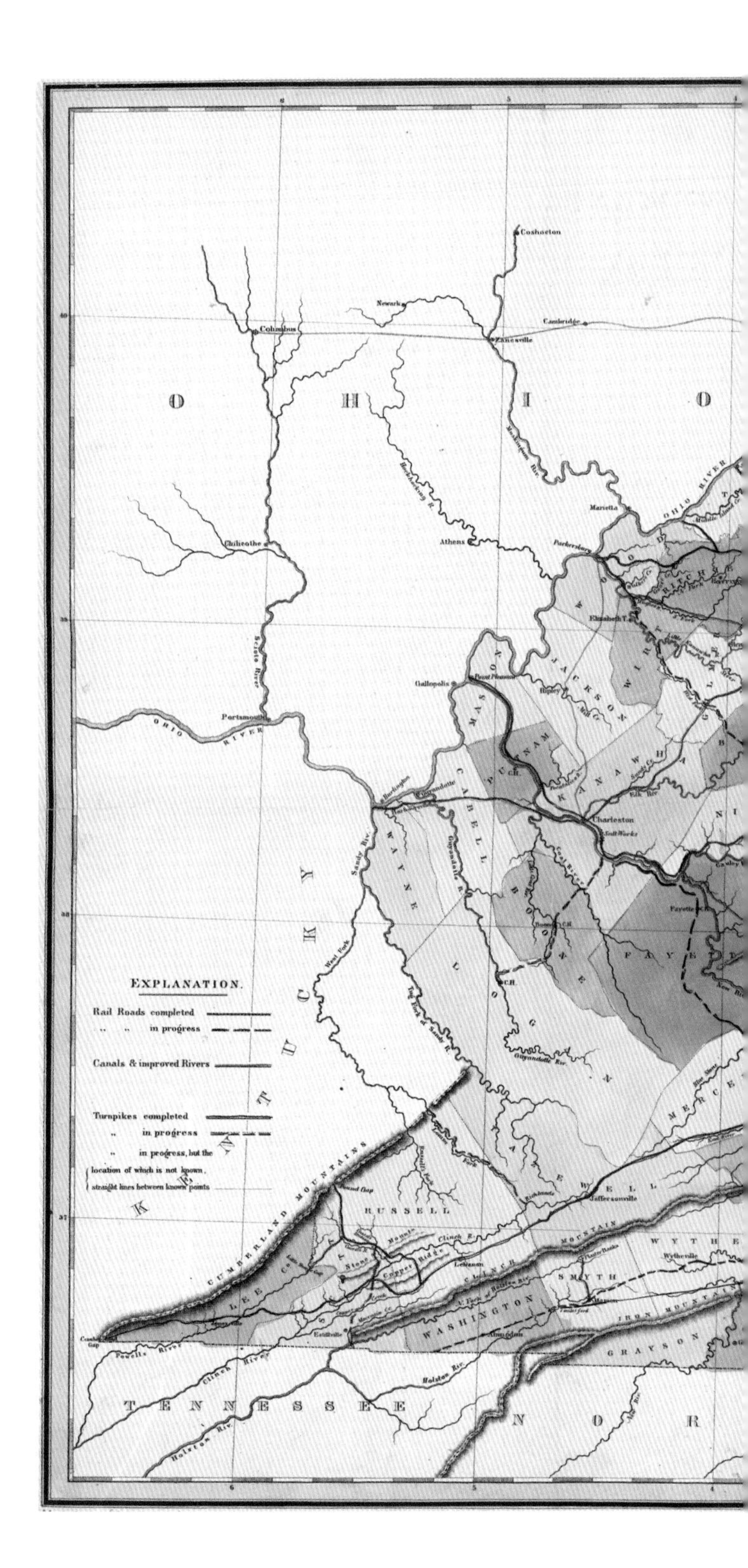

A map of the internal improvements of Virginia; prepared by C. Crozet, late principal engineer of Va. under a resolution of the General Assembly adopted March 15th 1848.

Cartographer Claudius Crozet helped found Virginia Military Institute and was elected principal engineer and surveyor for the Board of Public Works, supervising the building of canals and tunnels. By 1848 canal building had reached its peak; Virginia had built the James River and Kanawha Canal from Richmond to present-day West Virginia. Crozet oversaw the building of a railroad tunnel under the Blue Ridge, which at 4,273 feet was one of the longest in the world. It was used from 1858 to 1944.

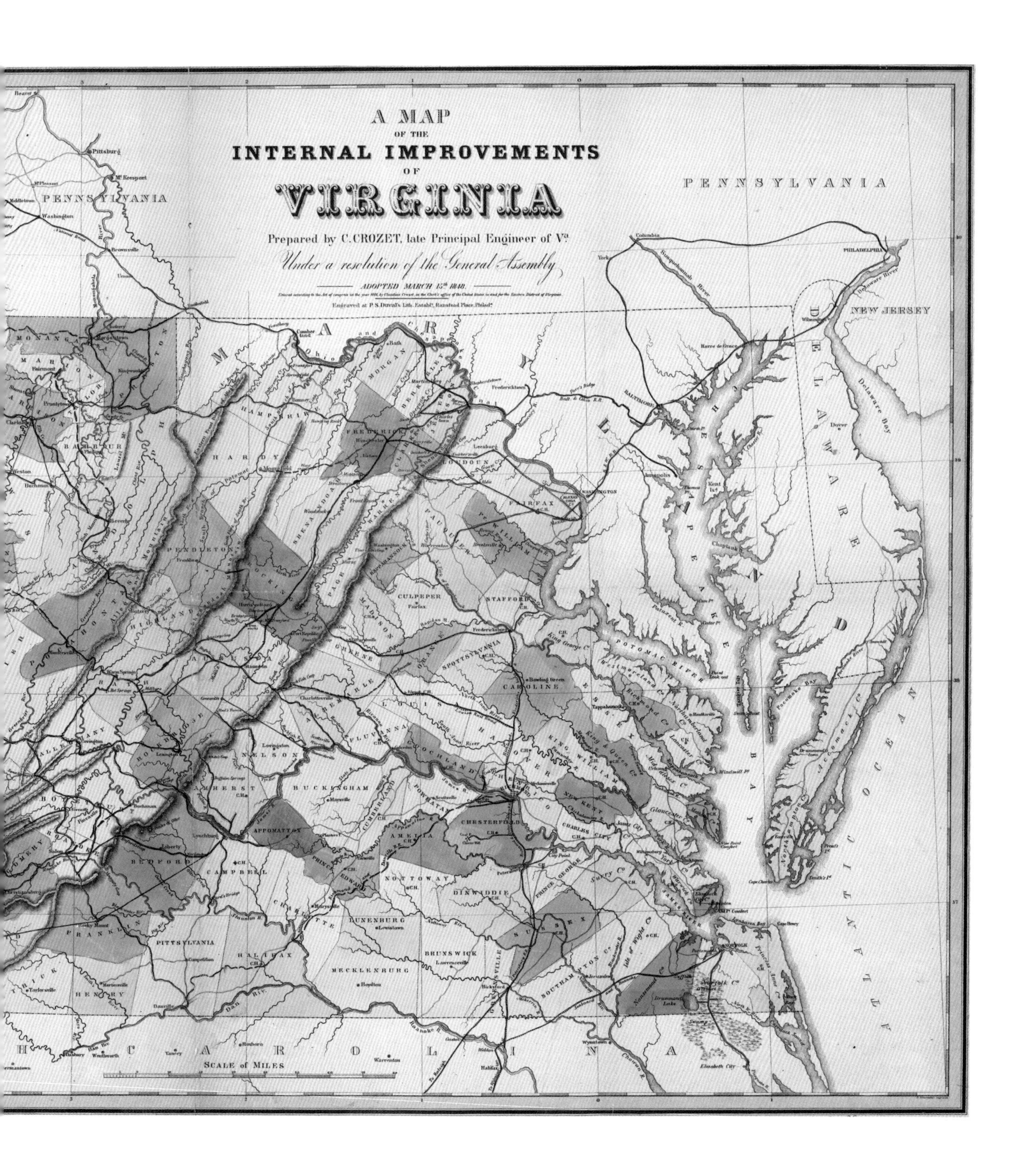
A MAP
OF THE
INTERNAL IMPROVEMENTS
OF
VIRGINIA
Prepared by C. CROZET, late Principal Engineer of Va.
Under a resolution of the General Assembly
ADOPTED MARCH 15th 1848.
Engraved at P. S. Duval's Lith. Establt. Ranstead Place, Philada.
PENNSYLVANIA
NEW JERSEY
DELAWARE
MARYLAND
POTOMAC RIVER
CHESAPEAKE BAY
ATLANTIC OCEAN
CAROLINA
SCALE of MILES

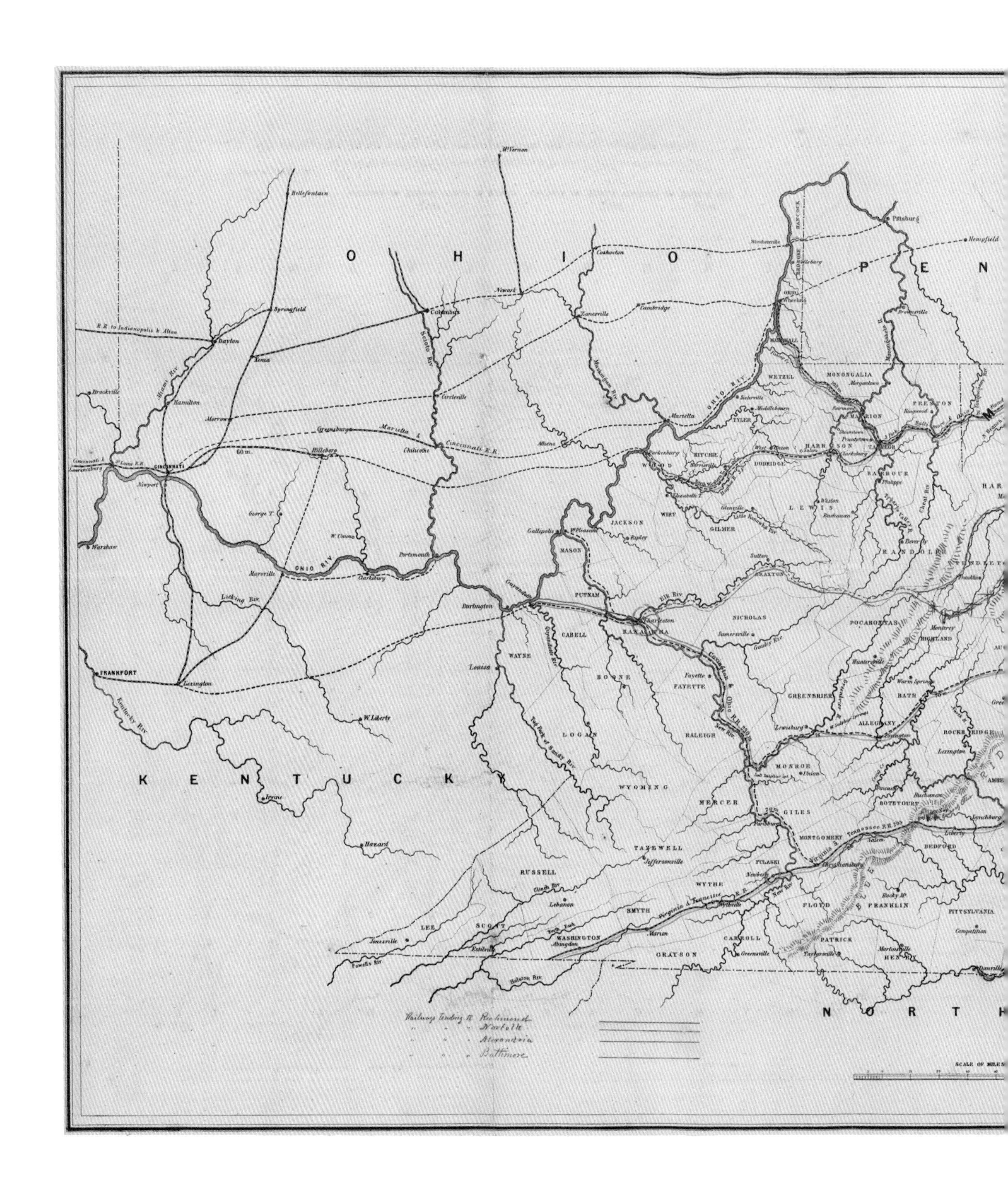

O H I O
P E N
K E N T U C K Y
N O R T H
Mt Vernon
Bellefontain
Coshocton
Newark
Columbus
Zanesville
Cambridge
Springfield
R.R. to Indianapolis & Alton
Dayton
Xenia
Brookville
Hamilton
Miami Riv
Scioto Riv
Circleville
Marietta
Athens
Chilicothe
Cincinnati R.R.
CINCINNATI
Newport
Hillsboro
60 m.
Ohio Riv
Portsmouth
Maysville
Licking Riv
FRANKFORT
Lexington
Kentucky Riv
W. Liberty
Irvine
Hazard
Louisa
Pittsburg
Wheeling
Parkersburg
Gallipolis
Burlington
Charleston
Elk Riv
New Riv
WAYNE
CABELL
PUTNAM
MASON
JACKSON
WIRT
RITCHIE
DODDRIDGE
GILMER
LEWIS
TYLER
WETZEL
MARSHALL
MONONGALIA
PRESTON
TAYLOR
BARBOUR
RANDOLPH
BRAXTON
NICHOLAS
KANAWHA
FAYETTE
BOONE
LOGAN
RALEIGH
WYOMING
MERCER
MONROE
GREENBRIER
POCAHONTAS
HIGHLAND
BATH
ALLEGHANY
GILES
BOTETOURT
MONTGOMERY
PULASKI
WYTHE
TAZEWELL
RUSSELL
SMYTH
WASHINGTON
LEE
SCOTT
GRAYSON
CARROLL
FLOYD
FRANKLIN
PATRICK
PITTSYLVANIA
BEDFORD
Lewisburg
Salem
Christiansburg
Lynchburg
Jeffersonville
Wytheville
Newbern
Marion
Abingdon
Lebanon
Jonesville
Estillville
Clinch Riv
Holston Riv
Virginia & Tennessee R.R.
Rocky Mt
Martinsville
Danville
Competition
Railways tending to Richmond
" " Norfolk
" " Alexandria
" " Baltimore
SCALE OF MILES

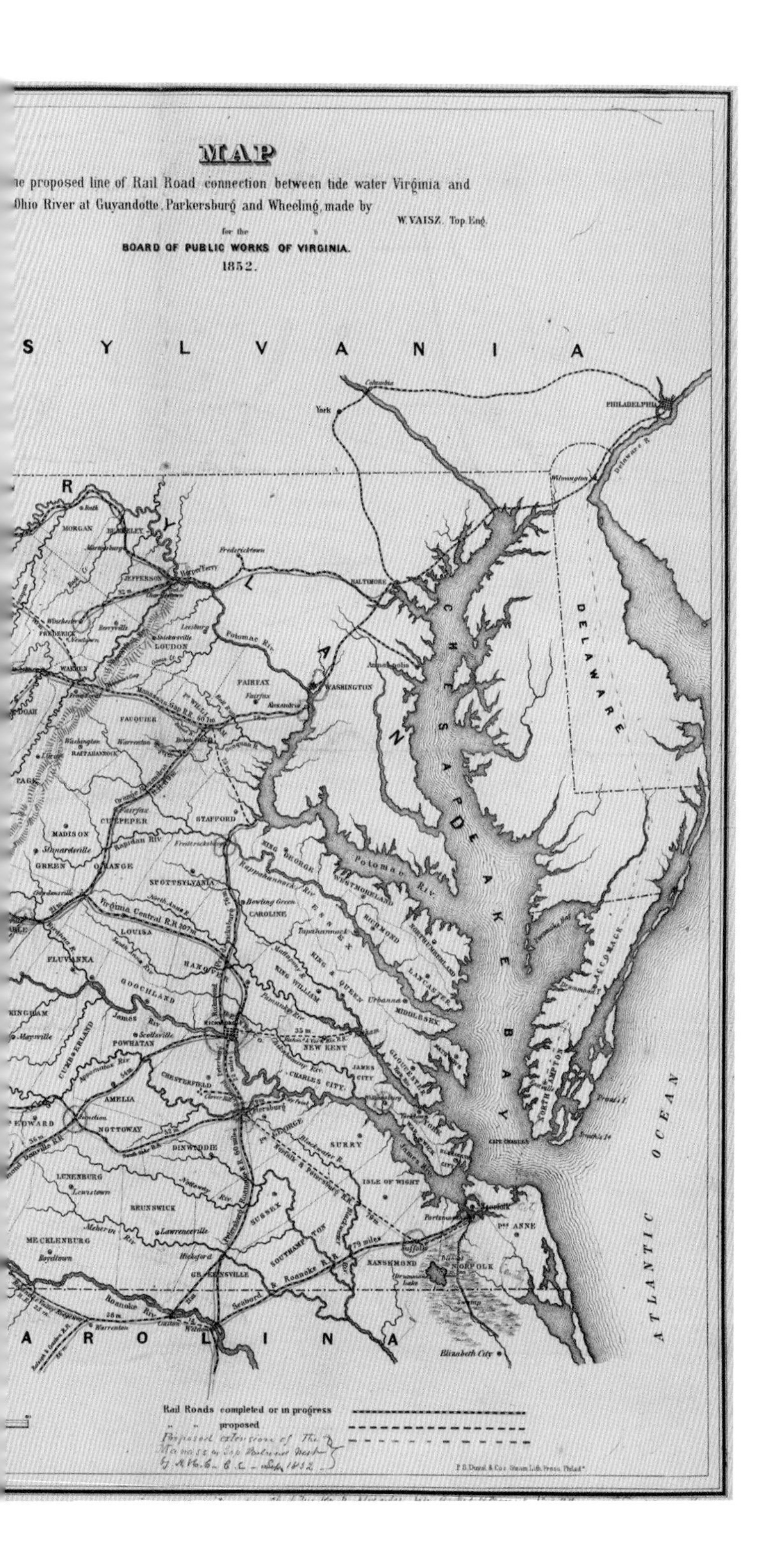

Map of the proposed line of Rail Road connection between tide water Virginia and the Ohio River at Guyandotte, Parkersburg and Wheeling, made by W. Vaisz, top. eng. for the Board of Public Works of Virginia.

Virginia had missed her chance to build a canal running westward from Alexandria on the Potomac, connecting Virginia's western territory with the Chesapeake Bay and the port of Norfolk. Maryland pushed ahead with a canal, and the C&O railroad later followed the canal route.

Plan of Rebel redoubt and barracks [at] "Camp Misery": surveyed April 11th 1862. Sneden, Robert Knox, 1832–1918. Courtesy of Virginia Historical Society.

A drawing of Confederate redoubt and barracks at "Camp Misery" on Ship Point Road near Yorktown. The Confederates evacuated the fort on April 5, 1862, as the siege of Yorktown began, and the drawing was made six days later by Robert Knox Sneden. A Brooklynite, he enlisted in the Army of the Potomac in 1861 and was captured by Mosby's troops in 1863. The "abatis" that he labels as surrounding the barracks are made up of felled trees with the smaller branches cut off.

The Civil War

By the time of the Civil War, secession of individual states from the United States had been suggested for decades, first by the small New England states, which thought that they were losing influence as population increased to the west. They saw that their interests lay in shipping and manufacturing, and western settlers would not be buying from them or shipping through their ports.

When the southern states threatened to secede, it was partly for the same reason. They saw the North manufacturing goods and voting high tariffs to keep out foreign goods. If the South bought foreign products, the tariff was high, and tariffs then were the major means of support for the Federal government. The resulting income was used in the west, not for Virginia's benefit. The South saw itself inevitably being outvoted. Its major assets were land and slaves. If the slaves were freed, the land could not be farmed and would become worth much less.

Zachary Taylor, a Tennessee slaveholder born in Virginia, won the election for the presidency in 1848, and the South thought that he would be sympathetic to their cause. However, he said that New Mexico and California could enter the union as free states, and that if any southern states tried to secede, he would stop them with force. Congress compromised on the issue: California was to be a free state, Utah and New Mexico Territories were open to slavery, the slave trade in D.C. was ended, and the Fugitive Slave Law was strengthened. The crisis passed for a time.

In the election of 1860, Abraham Lincoln ran for president, representing the new Republican Party. The opposition split. Democrats nominated Stephen Douglas, the Southern Democrats nominated John Breckenbridge of Kentucky, and the Constitutional Union Party nominated John Bell of Tennessee. When the news spread that Lincoln had won, South Carolina announced it would secede. President James Buchanan had his secretary of war, former Virginia governor John Floyd, write to the commander of Fort Sumter to order him to avoid hostile actions. Instead the commander moved troops from Fort Moultrie into Fort Sumter—a move considered aggressive by South Carolina. Floyd resigned when Buchanan refused to order the commander to return the troops to Fort Moultrie and instead ordered supplies sent to the fort.

South Carolina seceded, and its troops fired

on Fort Sumter in April 1861. A Virginian, Edmund Ruffin, asked for the privilege of firing the first shot, and it was granted. The South had committed an act of war. President Lincoln issued a call for troops. Virginia refused, and on April 17, 1861, it repealed its ratification of the Constitution and passed an act of secession. Richmond was chosen as the capital of the Confederate States of America, and Jefferson Davis was elected its president.

Robert E. Lee, who had served in the Mexican War with Ulysses Grant and Jefferson Davis, was offered the rank of colonel and command of Federal forces. Instead, he resigned and, as he said, took up his sword to defend his native state. He was made a major general and put in charge of the largest body of Confederate troops, the Army of Northern Virginia. Lee prevented Virginia from being invaded across the Potomac, and the first battles of the war were a rout for the North, but the outcome of the war was a foregone conclusion. The South had no navy to protect its ports, and therefore it was cut off from trade and help from other nations. It had few railroads and little manufacturing. Its farms had produced cotton and tobacco and could not turn easily to growing food for the troops. The North's population was greater, and it had access to the U.S. Treasury.

Meanwhile, Union armies in Virginia's former Ohio Territory marched southward and occupied the Virginia counties west of the Allegheny Mountains by early 1862. The following year, those counties were accepted into the union as the state of West Virginia. Virginia lost a third of its population and a third of its land area.

In March 1862, Virginia rebuilt the steamship *Merrimac* into the strange-looking ironclad vessel they renamed *Virginia.* It steamed across the river from Portsmouth to Hampton and sank two Union battleships, attempting to take control of the Chesapeake. However, the next morning a Union ironclad, the *Monitor,* arrived, and the two exchanged shots before both withdrew, neither winning decisively. The Chesapeake Bay and the James River, and the cities of Norfolk and Portsmouth, remained in Union control for the remainder of the war, making it easy for Union troops to reach Richmond.

Richmond was threatened with invasion in 1862, but Thomas "Stonewall" Jackson attacked from the west and forced Union troops back across the Potomac. This time the Union feared its capital might be invaded.

Lee attacked the North at Antietam, Maryland, one of the bloodiest battles ever fought on American soil, with heavy losses by both armies. Lee was defeated and retreated back into Virginia. Lincoln used this slight victory to issue his Emancipation Proclamation, freeing slaves in states still in rebellion. This changed nothing about the slaves' condition; however, it signaled to Britain and France that the Union was committed to ending slavery, not just saving the union. Those two countries then had a moral obligation to cease trade with the South.

For the next two years, Union forces tried unsuccessfully to conquer Virginia. Meanwhile, General Grant won victories along the Mississippi, cutting the Confederacy in two. Sherman struck eastward, across Georgia, devastating that state. Other Union forces occupied and burned the Valley of Virginia, destroying the South's major food supply. The noose was closing about Virginia, but there were still successes, such as Col.

John S. Mosby's capture of a train near Harpers Ferry that contained $173,000.

Lee struck northward again, in what would be a disastrous battle. His Army of Northern Virginia met the Union forces of Gen. George Meade near Gettysburg, Pennsylvania, in early July 1863. The Union army dug in atop Cemetery Ridge, and three dreadful days of fighting ended in fifty thousand casualties, Union and Confederate combined. The most slaughter occurred when Gen. George Pickett charged across an open field into direct fire. Lee would not attempt any further invasion of the North.

Lee was at his worst at Gettysburg and at his best at the Battle of the Wilderness in early May 1864, managing to keep Grant away from Richmond. Again at Cold Harbor in June, Lee held Virginia territory, and Grant's army suffered huge losses.

Southerners hoped that Lincoln would lose the election of 1864 and a new president would negotiate a fair end to the war, but Lincoln won. Confederate armies were so depleted that in May 1864 Virginia Military Institute cadets, ages fourteen to eighteen, went into battle at New Market, Virginia.

In the summer of 1864, with Lee's army dug in around Petersburg, a Pennsylvanian suggested digging underneath the Confederate lines to blast a hole. Troops could then attack the Confederates from behind. The blast, called the Battle of the Crater, killed 250 Confederates in an instant, but before the day ended, 3,798 Union soldiers had died, many trapped and buried in debris, or shot as they attempted to climb out of the crater.

In late March 1865, Lee decided to join forces with troops in North Carolina. His army was weary and short of supplies, and on their retreat to the southwest, their wagon trains were seized by Grant's pursuing army. Lee had no choice but to surrender.

The two generals, Lee and Grant, met in Wilbur McLean's house in Appomattox, Virginia, on April 9 to agree to terms of surrender. Ironically, McLean had moved his family from northern Virginia after the First Battle of Manassas in 1861, seeking safety in the central Virginia village.

The long, hopeless war was over.

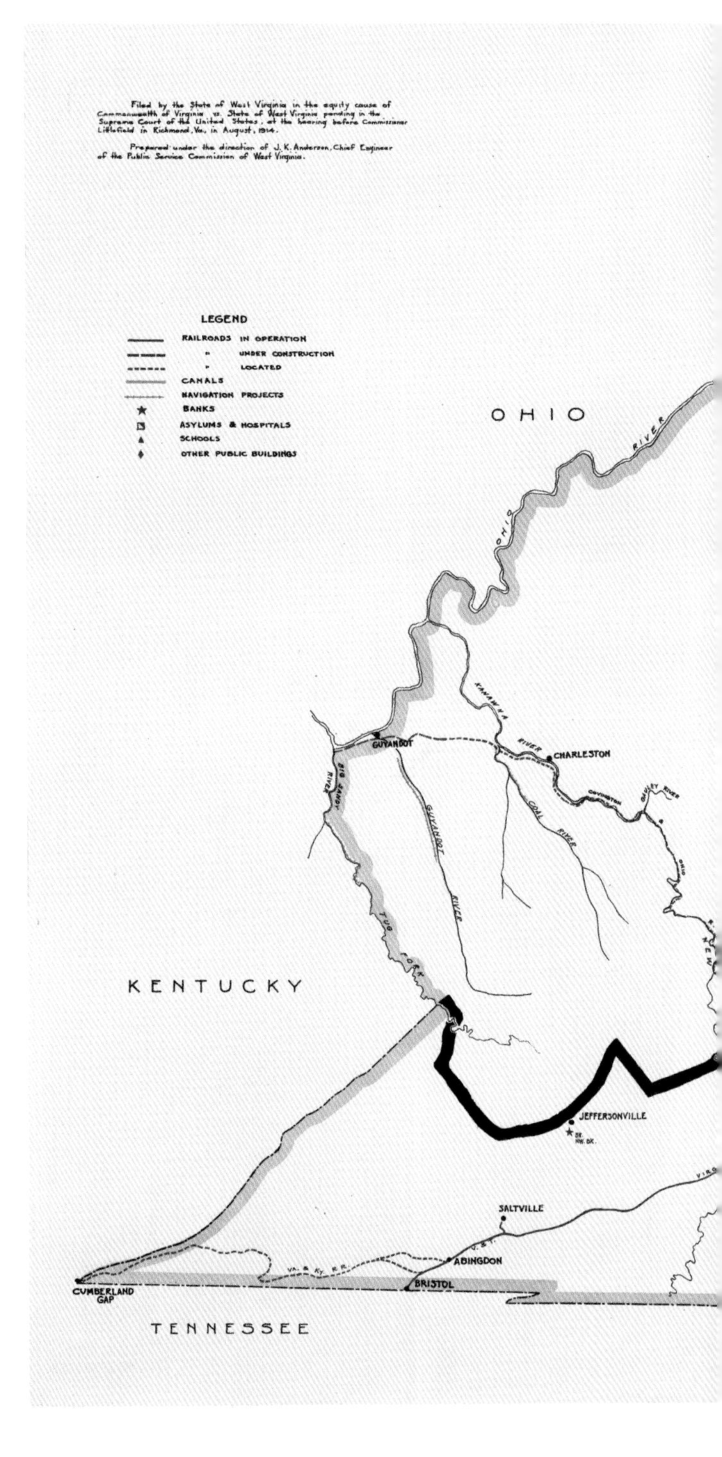

Map showing the location of railroads, canals, navigation projects and public institutions in which the Commonwealth of Virginia had invested money as of date January 1st. 1861: as traced from an official map in the possession of the Virginia State Library entittled [sic] "A map of the State of Virginia reduced from the Nine Sheet Map of the State in conformity to law by Herman Boye, 1828, corrected by order of the Executive 1859 by L.v. Buckholtz": together with the division line later established between Virginia and West Virginia and additional extensions made from completion of map until January 1st, 1861. Courtesy of the Library of Virginia.

This map showing railroads, canals, navigation projects, and public institutions that Virginia had invested money in as of January 1, 1861, and for which the state had assumed debt, is an updated version of the 1828 Nine Sheet Map prepared by Herman Böÿe (see page 48). This map was part of the lawsuit Virginia filed against West Virginia after the end of the Civil War, claiming that the breakaway counties owed part of the debt, although most of the improvements (shown as red lines) were in the remaining Old Dominion.

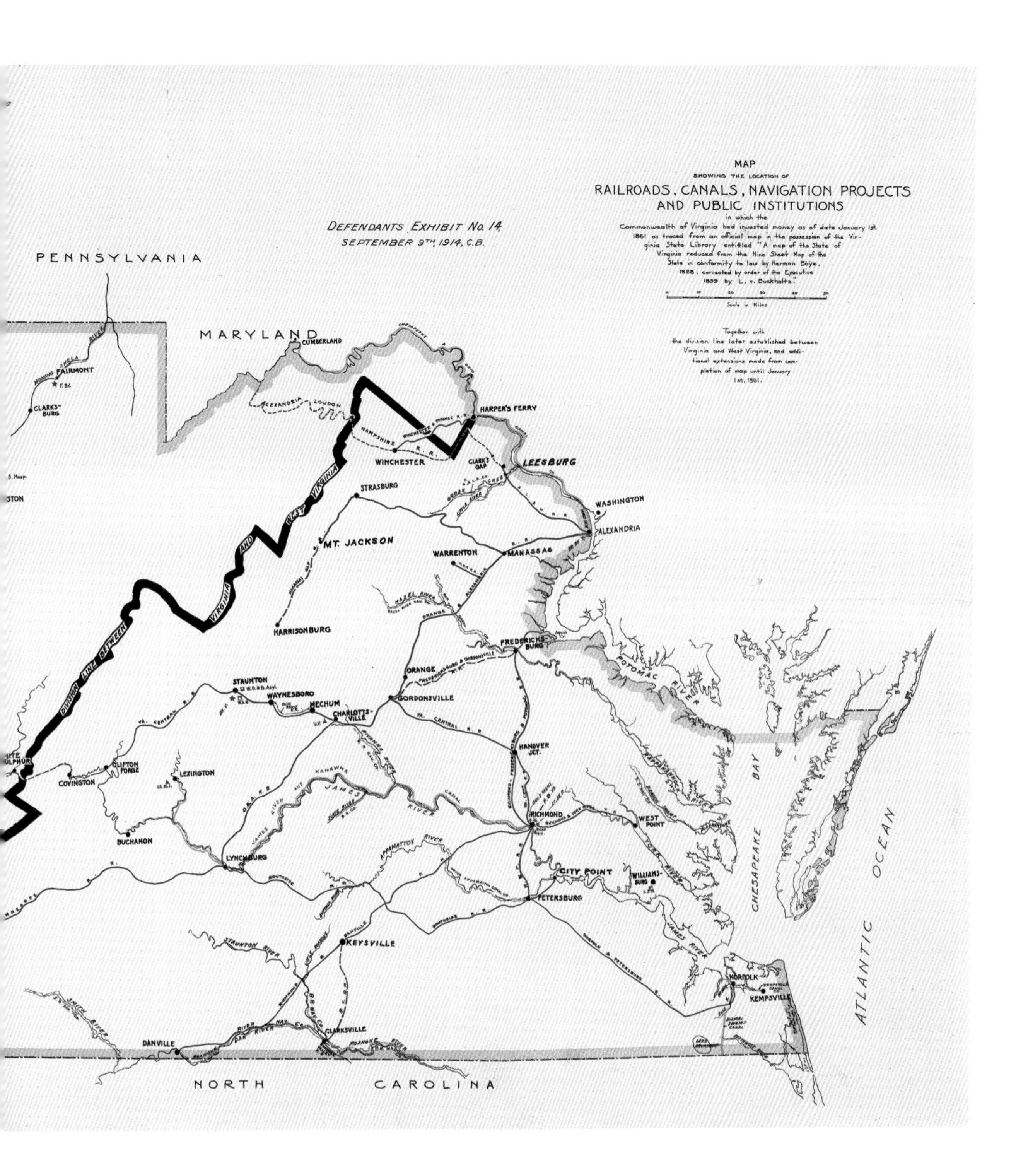
MAP
SHOWING THE LOCATION OF
RAILROADS, CANALS, NAVIGATION PROJECTS
AND PUBLIC INSTITUTIONS
in which the
Commonwealth of Virginia had invested money as of date January 1st
1861 as traced from an official map in the possession of the Vir-
ginia State Library entitled "A map of the State of
Virginia reduced from the Nine Sheet Map of the
State in conformity to law by Herman Böye,
1828, corrected by order of the Executive
1859 by L. v. Buckholtz."
0 10 20 30 40 50
Scale in Miles
Together with
the division line later established between
Virginia and West Virginia, and addi-
tional extensions made from com-
pletion of map until January
1st, 1861.
DEFENDANTS EXHIBIT No. 14,
SEPTEMBER 9TH, 1914, C.B.
PENNSYLVANIA
MARYLAND
CUMBERLAND
FAIRMONT
CLARKSBURG
HARPER'S FERRY
WINCHESTER
LEESBURG
STRASBURG
WASHINGTON
ALEXANDRIA
MT. JACKSON
WARRENTON
MANASSAS
HARRISONBURG
FREDERICKSBURG
STAUNTON
WAYNESBORO
MECHUM
CHARLOTTESVILLE
ORANGE
GORDONSVILLE
HANOVER JCT.
COVINGTON
CLIFTON FORGE
LEXINGTON
BUCHANON
LYNCHBURG
RICHMOND
WEST POINT
CITY POINT
WILLIAMSBURG
PETERSBURG
KEYSVILLE
NORFOLK
KEMPSVILLE
DANVILLE
CLARKSVILLE
DIVISION LINE BETWEEN VIRGINIA AND WEST VIRGINIA
POTOMAC RIVER
JAMES RIVER
APPAMATTOX RIVER
STAUNTON RIVER
CHESAPEAKE BAY
ATLANTIC OCEAN
NORTH CAROLINA

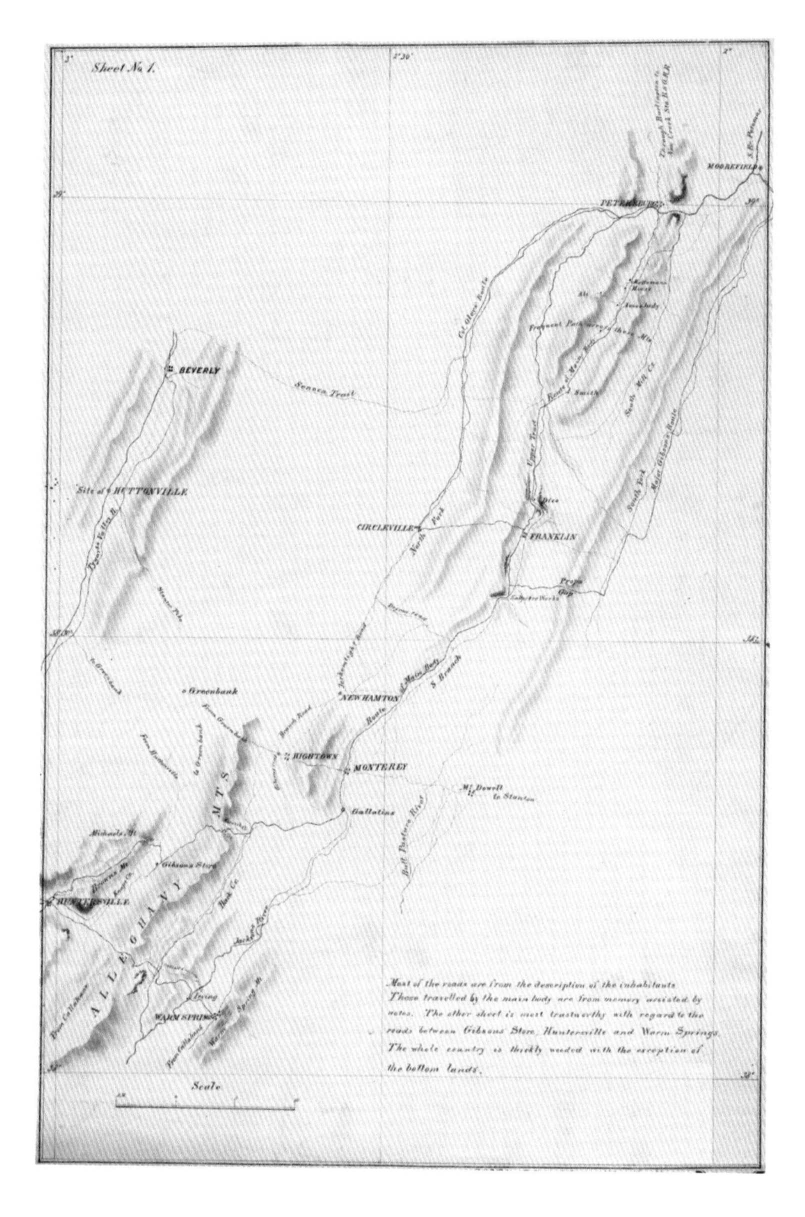

[Western Virginia from Petersburg to Warm Springs, showing the movement of the Union army, 1862], drawen [sic] by A. Hausmann.

Map showing the movement of Union troops across western Virginia in 1862. A line at the upper right shows "Frequent path across these mountains." A red line indicates "Route of the main body." Warm Springs, Virginia, is at the bottom left, and Petersburg, in western Virginia, is at the top right. The artist comments, "Most of the roads are from the descriptions of the inhabitants. Those traveled by the main body are from memory assisted by notes. The country is thickly wooded with the exception of the bottom lands." This made it easier for troops to hide.

Scene of the late naval fight and the environs of Fortress Monroe, and Norfolk and Suffolk, now threatened by General Burnside.

This wood block print is from William Jenning Demorest's *New York Illustrated News,* March 22, 1862. The scene is of Fortress Monroe, the Gosport Navy Yard at Portsmouth, and the *Monitor,* at the time of General Burnside's advance on Suffolk, Virginia. The map was originally oriented so that north was to the left (here, at the top), and Williamsburg was at the upper left (now upper right).

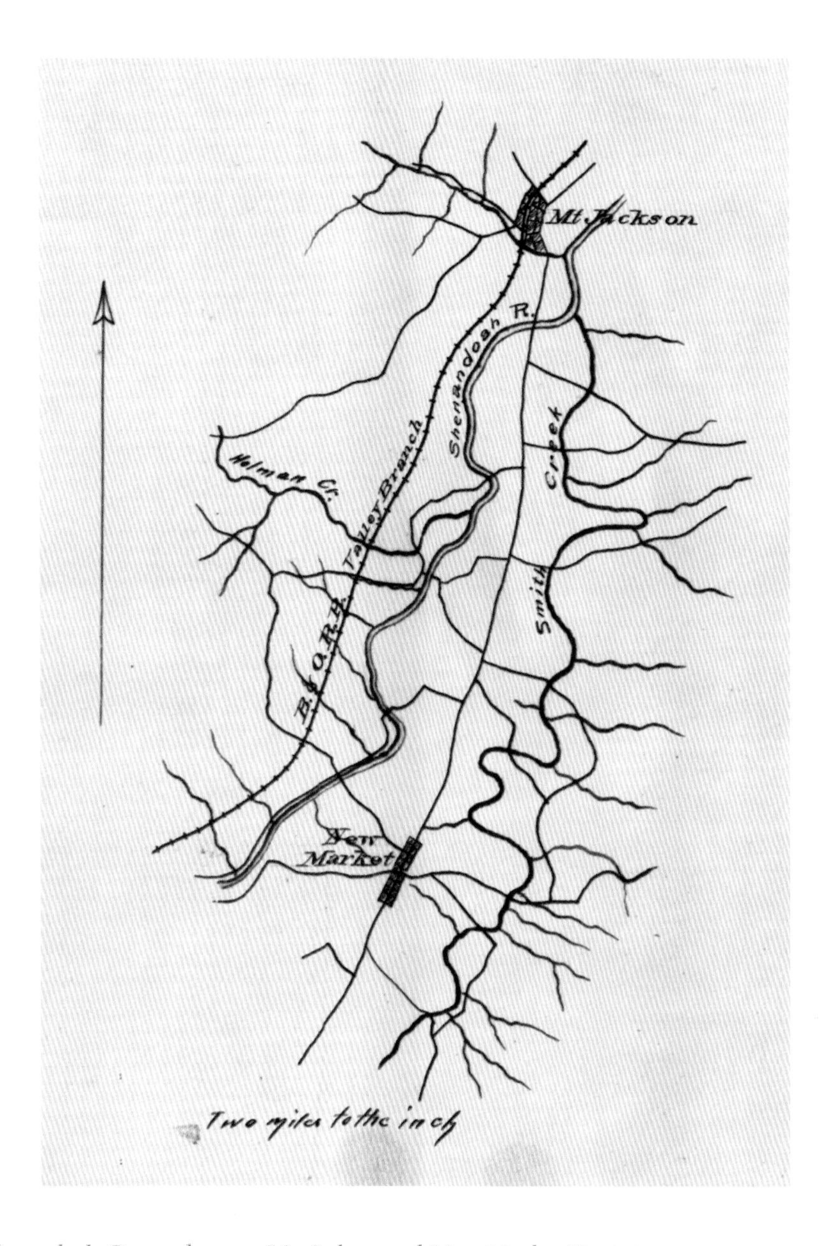

Map of Shenandoah County between Mt. Jackson and New Market, Virginia.

The Baltimore and Ohio Railroad is shown at the left, roughly paralleling the Shenandoah River. Gen. Franz Sigel was ordered to destroy the Valley railroads and marched south from Winchester. Maj. Gen. Cabell Breckenridge, along with 263 teenage cadets from Virginia Military Institute, marched northward and encountered Sigel at New Market. Five cadets were killed, five others died later from wounds, and forty-seven were wounded, but they captured several cannons and won a victory for the Confederates. Drawn in 1863, the map's cartographer is unknown.

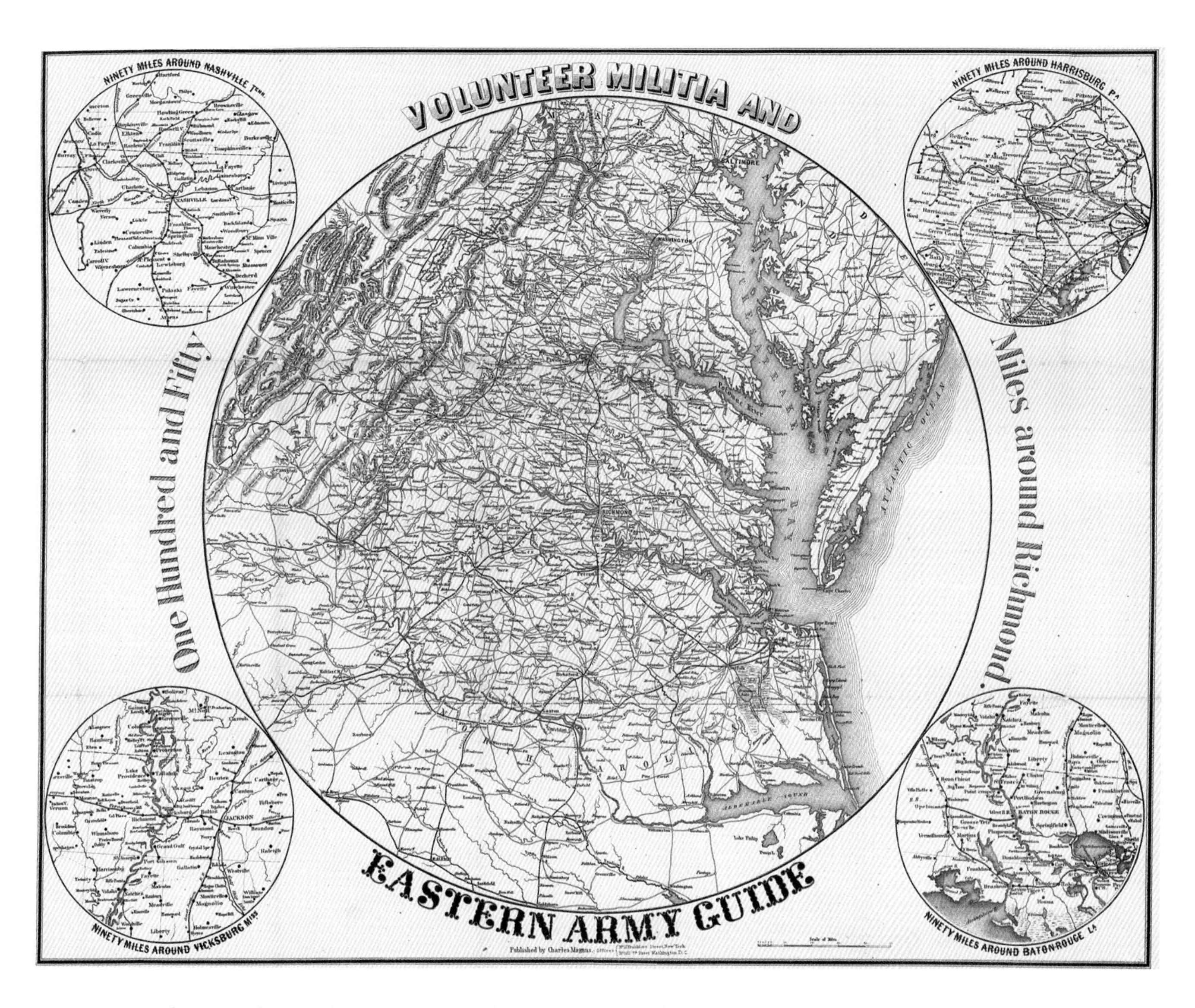

Volunteer militia and eastern army guide. One hundred and fifty miles around Richmond.
With this map, troops could find their way about the Old Dominion, for it details mountains, rivers, towns, roads, and railroads. The four insets show features within ninety miles around Nashville, Tennessee; Harrisburg, Pennsylvania; Vicksburg, Mississippi; and Baton Rouge, Louisiana.

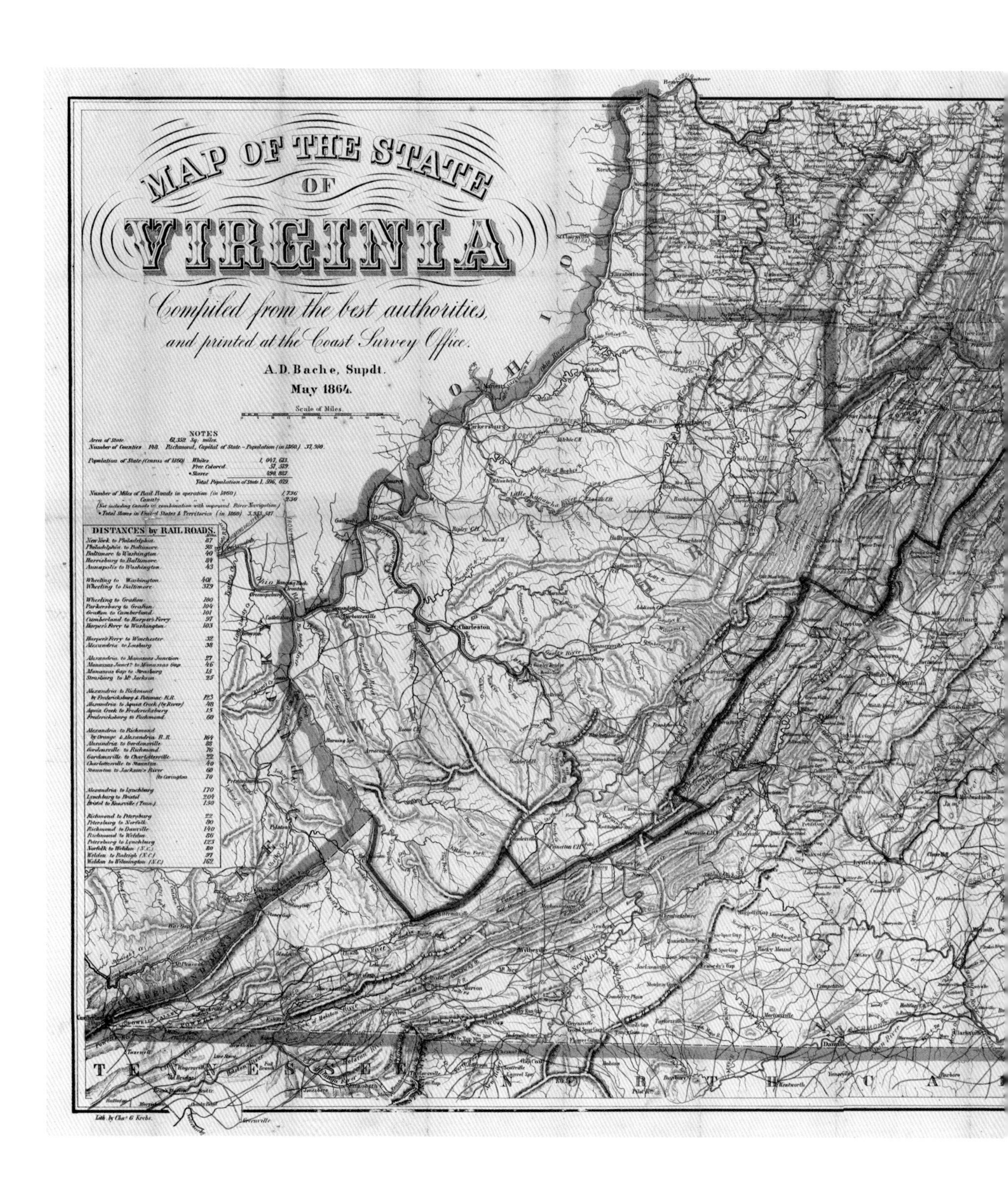
MAP OF THE STATE
OF
VIRGINIA
Compiled from the best authorities,
and printed at the Coast Survey Office.
A. D. Bache, Supdt.
May 1864.
Scale of Miles.
NOTES
Area of State 61,352 Sq: miles.
Number of Counties 148. Richmond, Capital of State – Population (in 1860.) 37,910.
Population of State (Census of 1860) Whites 1,047,613.
Free Colored 57,579.
*Slaves 490,887.
Total Population of State 1,596,079.
Number of Miles of Rail Roads in operation (in 1860) 1,736
Canals 230
(Not including Canals in combination with improved River Navigation.)
*Total Slaves in United States & Territories (in 1860) 3,953,587
DISTANCES by RAIL ROADS.
Miles
New York to Philadelphia 87
Philadelphia to Baltimore 98
Baltimore to Washington 40
Harrisburg to Baltimore 84
Annapolis to Washington 43
Wheeling to Washington 401
Wheeling to Baltimore 379
Wheeling to Grafton 100
Parkersburg to Grafton 104
Grafton to Cumberland 101
Cumberland to Harper's Ferry 97
Harper's Ferry to Washington 103
Harper's Ferry to Winchester 32
Alexandria to Leesburg 38
Alexandria to Manassas Junction 27
Manassas Junct. to Manassas Gap 46
Manassas Gap to Strasburg 15
Strasburg to Mt. Jackson 25
Alexandria to Richmond
by Fredericksburg & Potomac R.R. 123
Alexandria to Aquia Creek (by River) 48
Aquia Creek to Fredericksburg 15
Fredericksburg to Richmond 60
Alexandria to Richmond
by Orange & Alexandria R.R. 164
Alexandria to Gordonsville 88
Gordonsville to Richmond 76
Gordonsville to Charlottesville 22
Charlottesville to Staunton 40
Staunton to Jackson's River 60
(to Covington 70
Alexandria to Lynchburg 170
Lynchburg to Bristol 204
Bristol to Knoxville (Tenn.) 130
Richmond to Petersburg 22
Petersburg to Norfolk 80
Richmond to Danville 140
Richmond to Weldon 86
Petersburg to Lynchburg 123
Norfolk to Weldon (N.C.) 80
Weldon to Raleigh (N.C.) 97
Weldon to Wilmington (N.C.) 162
O H I O
Charleston
Parkersburg
Lynchburg
Danville
Wytheville
T E N N E S S E E
N O R T H C A
Lith. by Chas. G. Krebs.

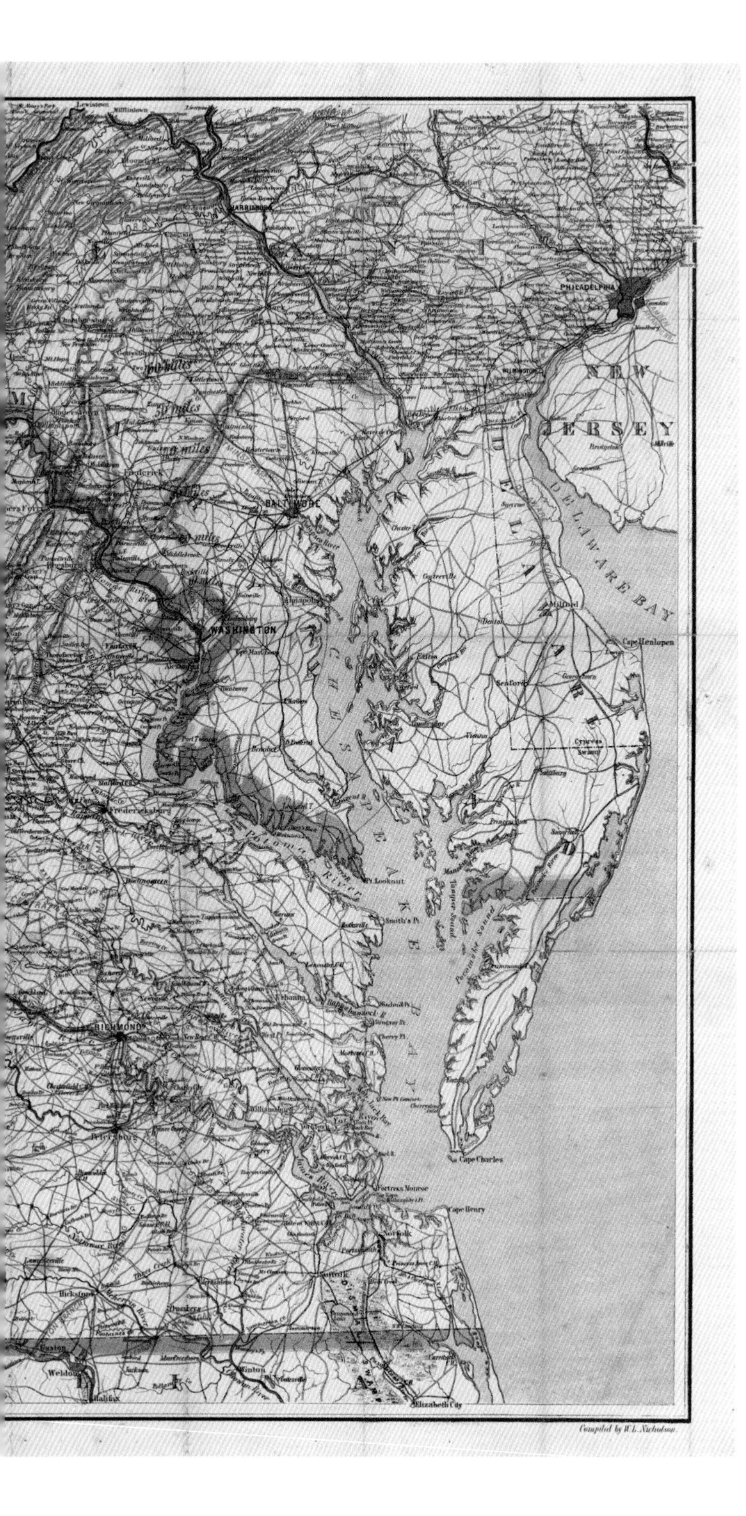

Map of the State of Virginia compiled from the best authorities, and printed at the Coast Survey Office. A. D. Bache, Supdt. May 1864. Compiled by W. L. Nicholson.

Red lines indicate railroads: the Manassas Gap going toward Alexandria; Virginia Central through Rockfish Gap; Orange & Alexandria from Charlottesville; Richmond, Petersburg & Potomac going north–south; the City Point & York River going down the peninsula; and the Norfolk & Petersburg, which became a major coal carrier. Mileage between points is given at the left.

Map of Henrico County, Va.: showing fortifications around Richmond, north and east of the James River.

This map of Henrico County, Virginia, shows the Confederate fortifications around Richmond north and east of the James River, as drawn in 1864 by an unknown cartographer. The names of many landowners are listed, and indications are that nearly all had fortified their homes in some way. The small tear-shaped peninsula at the bottom right is Curle's Neck, the same bit of land Sir Thomas Dale had chosen for a fortress in 1610.

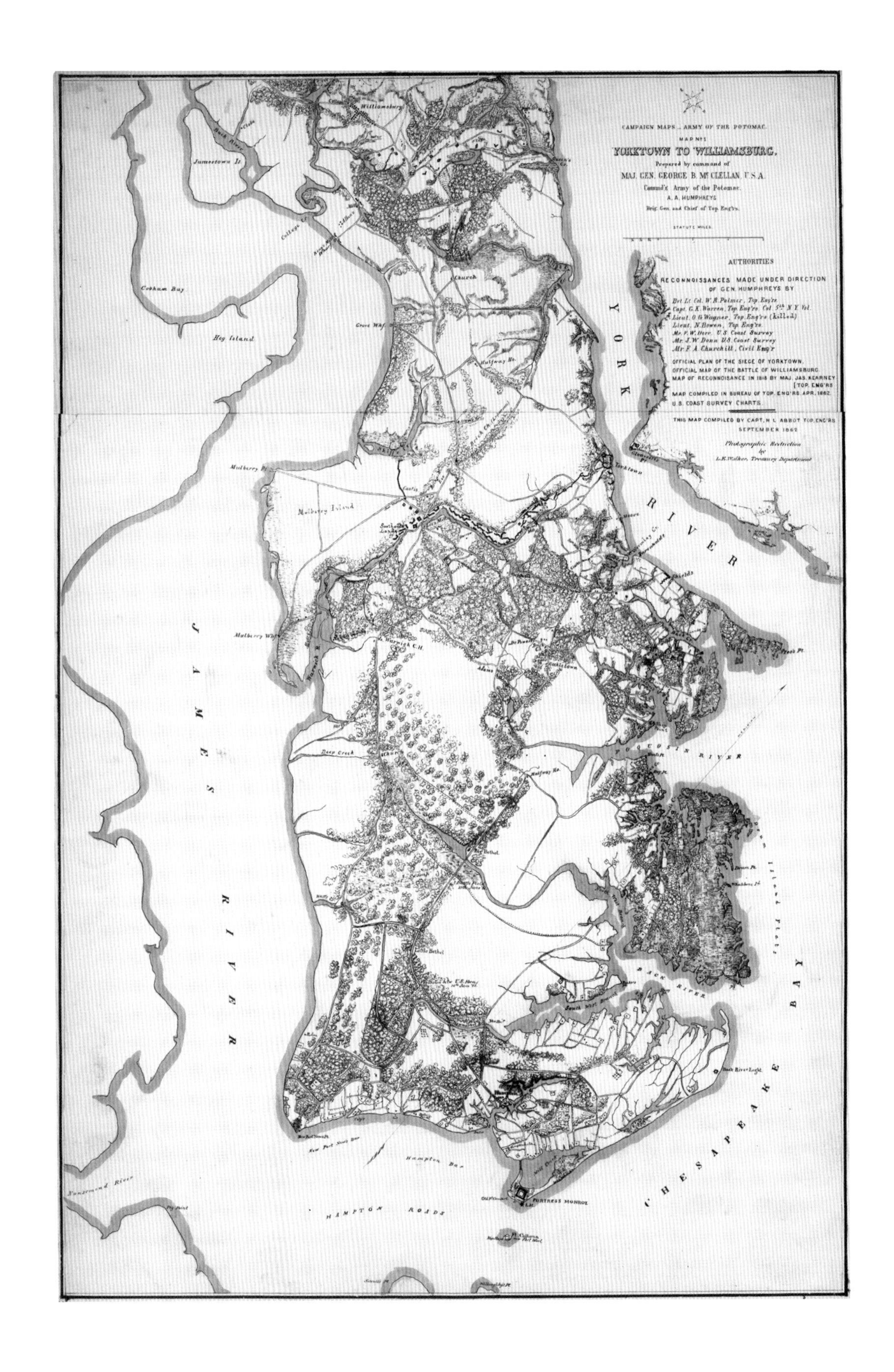
CAMPAIGN MAPS _ ARMY OF THE POTOMAC.
MAP No. 1
YORKTOWN TO WILLIAMSBURG,
Prepared by command of
MAJ. GEN. GEORGE B. McCLELLAN, U.S.A.
Command'g Army of the Potomac.
A. A. HUMPHREYS
Brig. Gen. and Chief of Top. Eng'rs.
STATUTE MILES.
AUTHORITIES
RECONNOISSANCES MADE UNDER DIRECTION OF GEN. HUMPHREYS BY
Brt. Lt. Col. W. R. Palmer, Top. Eng'rs.
Capt. G. K. Warren, Top. Eng'rs. Col. 5th N. Y. Vol.
Lieut. O. G. Wagner, Top. Eng'rs (killed)
Lieut. N. Bowen, Top. Eng'rs.
Mr. F. W. Dorr. U.S. Coast Survey
Mr. J. W. Donn U.S. Coast Survey
Mr. F. A. Churchill, Civil Eng'r
OFFICIAL PLAN OF THE SIEGE OF YORKTOWN.
OFFICIAL MAP OF THE BATTLE OF WILLIAMSBURG.
MAP OF RECONNOISANCE IN 1818 BY MAJ. JAS. KEARNEY [TOP. ENG'RS
MAP COMPILED IN BUREAU OF TOP. ENG'RS. APR. 1862.
U.S. COAST SURVEY CHARTS.
THIS MAP COMPILED BY CAPT. H. L. ABBOT TOP. ENG'RS
SEPTEMBER 1862
Photographic Restitution by L.E. Walker, Treasury Department
Williamsburg
Jamestown Id.
Cobham Bay
Hog Island
Mulberry Island
Mulberry Pt.
Mulberry Whf.
Deep Creek
Warwick C.H.
Yorktown
YORK RIVER
JAMES RIVER
POQUOSIN RIVER
BACK RIVER
CHESAPEAKE BAY
Hampton Bar
HAMPTON ROADS
FORTRESS MONROE
Nansemond River
Church
Grove Whf.
Halfway Ho.
Bethel
Little Bethel

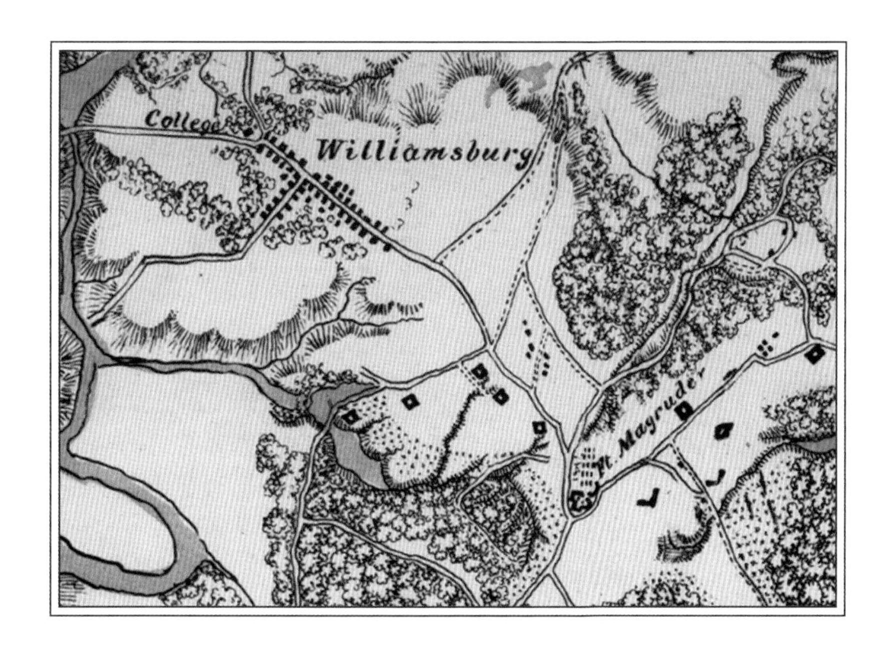

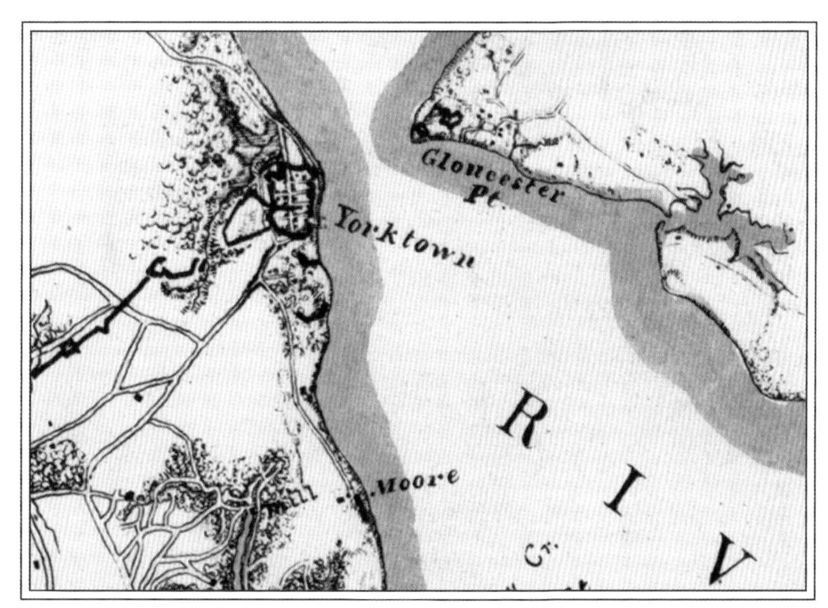

Yorktown to Williamsburg. This map compiled by Capt. H. L. Abbot, Top. Eng'rs., September 1862. Photographic reduction by L. E. Walker, Treasury Department.

Map number 1 of the Campaign Maps of the Army of the Potomac, the official plan of the siege of Yorktown and the official map of the Battle of Williamsburg. Compiler Capt. H. L. Abbot, at the command of Maj. Gen. George B. McClellan, oriented the map so that north is to the right, the James River is at the left, and the Chesapeake Bay is at the bottom. The detailed map of Williamsburg shows the colonial capital of Virginia and a possible route to Richmond, capital of the Confederacy. The detail of Yorktown shows its location at a narrow point of the York River. Control of the town would mean control of the river traffic on the peninsulas leading to Richmond.

1 Hunting Creek Bridge to Mount Vernon 2 Fort Lyon
3 Fort Ellsworth 4 Fairfax Seminary
8 Provost Marshal's Office

BIRDS EYE VIEW OF ALEXAN

Birds eye view of Alexandria, Va.

At the time this was drawn, in 1863 by Charles Magnus, Alexandria was a large, important port. Just across the river from Washington, D.C., it had early been seized by Union forces, and all the boats in the harbor in this drawing are flying Union flags. A train at center is bringing more goods to port.

Map of Appomattox Court House and vicinity. Showing the relative positions of the Confederate and Federal Armies at the time of General R. E. Lee's surrender, April 9th 1865.
Lee's headquarters is at the upper center, indicated by a red line. Grant's headquarters is at the lower left, indicated by a blue line. The picture at the bottom left is of McLean House, site of the surrender, and at the bottom right, Lee says farewell to his troops. The other photos and notes show army tents, Grant's headquarters, where weapons were stacked, and local houses.

Lee's Head-Quarters.

Historical Notes.

On Sunday, the 2d of April 1865, General Lee was holding at Petersburg a semi-circular line south of the Appomattox River, with his left resting on the river, and his right on the South Side Rail Road, fifteen miles from the city. The Federals were pressing his whole line. Sheridan with his cavalry on the right. To save his right flank, General Lee telegraphed to Richmond, that during the night he would fall back to the north side of the river, and ordered that Richmond be evacuated simultaneously.

On the morning of the 3d the retreat commenced in earnest, General Grant hurrying up to get possession of Burkesville—the junction of the South Side and Danville Railroad—in hopes of cutting off General Lee from Danville or Lynchburg. On the 5th a portion of the Federal forces occupied Burkesville, Sheridan with his cavalry being in advance at Jetersville on the Danville Railroad. General Lee at Amelia C. H., 6 miles north of Sheridan's advance. In this situation General Sheridan telegraphed:—"I feel confident of capturing the entire Army of Northern Virginia, if we exert ourselves. I see no escape for Lee." On the evening of the 6th some heavy fighting took place between the Federal advance and Lee's retreating column. Sheridan again telegraphed: "If the thing is pressed I think Lee will surrender." Lee continued to press for Lynchburg—his men probably anticipating the result, daily leaving him by thousands,—until on the morning of the fated 9th of April, 1865, he confronted the overwhelming forces of Gen. Grant with a little less than 8,000 muskets.

The position of the Confederate army was briefly this: occupying the narrow strip of land between the South Side Railroad and the James River; the only road on which it was possible to retreat, was that marked Lynchburg road on the map. Sheridan with his cavalry having struck the railroad at Appomattox Station, obtaining possession of the Lynchburg road, thus effectually cutting off Lee's retreat. Gen. Lee now had the choice of either cutting his way directly through the Federal forces, or immediate surrender. In view of the immense disparity of forces between the ranks of the half starved Confederates and the overwhelming army of General Grant, he chose the letter alternative.

Generals Lee and Grant met at the house of Wilmer McLane, Esq., and after a brief interview, at 3½ o'clock p. m.

McLane's House.

GEN. LEE'S FA

HEAD-QU.

After four years of a
courage and fortitude,
been compelled to yie
sources.

I need not tell the
tles, who have remai
consented to this resul
that valor and devotio
compensate for the loss
ation of the contest,

198 Entered according to act of Congress, in the year 1866, by Henderson & Co. in the Clerk's office of

MAP OF
[AP]POMATTOX COURT HOUSE
AND VICINITY.

[Showing th]e relative positions of the Confederate and Federal Armies at the time of General R. E. LEE'S Surrender, April 9th 1865.

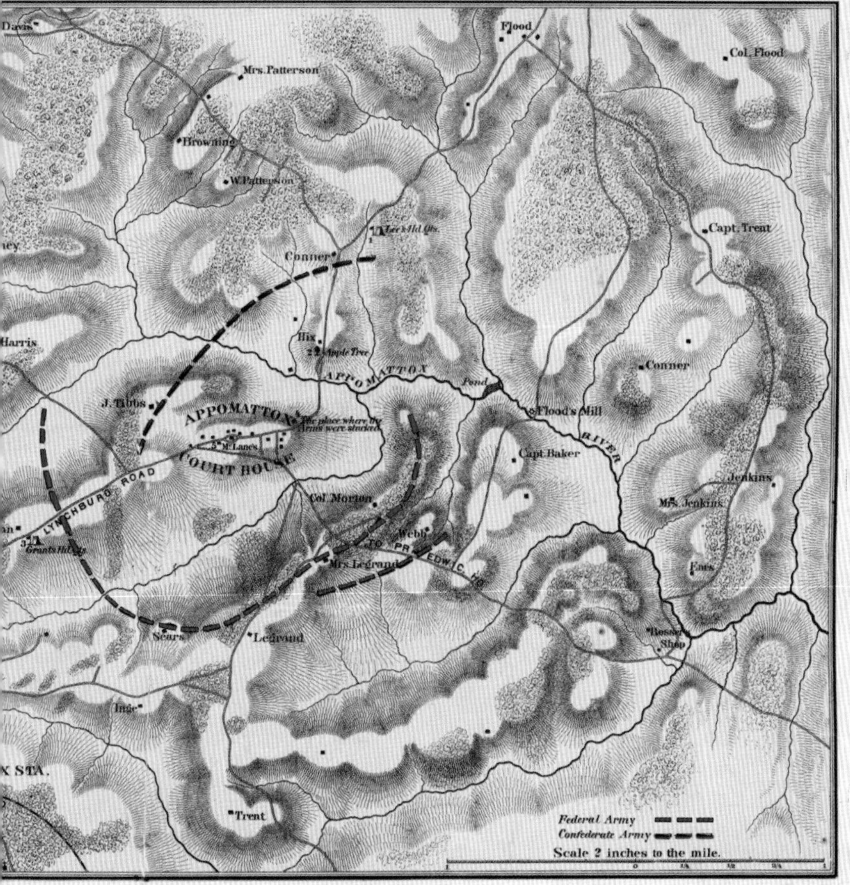

Grant's Head-Quarters.

on the 9th of April 1865, the Articles of Capitulation were signed by General Lee. While negotiations were being conducted by the two Commanders-in-Chief, the General officers of either army were mingling socially together in the streets of Appomattox C. H., and drinking mutual healths. Gens. Ord, Sheridan, Gibbon, Michie and others of the Federals, Gens. Longstreet, Heath, Gordon and others, of the Confederates.

At 4 o'clock p. m. the announcement of Lee's surrender was made to Grant's army. The wildest enthusiasm immediately broke forth, and all seemed mad with joy.

As the great Confederate General rode past his gallant little band from his interview with Gen. Grant, whole lines of battle rushed to the beloved old chief, and breaking ranks, each struggled with the other to wring him by the hand. With tears rolling down his cheeks, General Lee could only say, "Men, we have fought through the war together. I have done the best that I could for you."

On the morning of the 12th April the Army of Northern Virginia marched by divisions to a point near Appomattox Court House, and stacked arms and accoutrements. Maj. Gen. Gibbon representing the United States authorities.

On the afternoon of the 12th, with an escort of Federal cavalry as a guard of honor, attended by a portion of his staff, General Lee returned to Richmond.

Thus quietly passed from the theater of the most desperate war of modern times the renowned Commander of the Army of Northern Virginia, and the remnants of that once invincible army were quietly wending their way to their long forsaken homes.

LIST OF ENGRAVINGS.

Gen. Lee's Head-Quarters near Conner's House.—Position marked by a flag and No. 1 on the map.

View of Appomattox Court House.

General Grant's Head-Quarters near Coleman's House.—Position marked by a flag and No. 3 on the map.

Place where the arms were stacked. The exact spot is marked No. 4 on the map. In this picture may be seen the famous apple tree, (position marked with a tree and No. 2 on the map,) near Hix's house, where the first meeting between the Commanders was generally, but incorrectly, supposed to have taken place.

McLane's House, in the village of Appomattox Court House, where the articles of capitulation were signed. The signing took place in the front room, on the right of the door, entering from the porch.

[...]HIS ARMY.

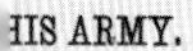

[...]N VIRGINIA,
1865.

[...]d by unsurpassed [...]rn Virginia has [...]umbers and re-

[...] hard-fought bat[...]st, that I have [...]em; but feeling [...]thing that could [...]ded the continu[...]void the useless

Appomattox Court House.

sacrifice of those whose past services have endeared them to their countrymen.

By the terms of agreement, officers and men can return to their homes and remain there until exchanged.

You will take with you the satisfaction that proceeds from the consciousness of duty faithfully performed; and I earnestly pray that a merciful God will extend to you His blessing and protection.

With an unceasing admiration of your constancy and devotion to your country, and a grateful remembrance of your kind and generous consideration of myself, I bid you an affectionate farewell

R. E. LEE, *General.*

Place where the Arms were Stacked.

LITH. by A. HOEN & CO. Baltimore.

[...]land.

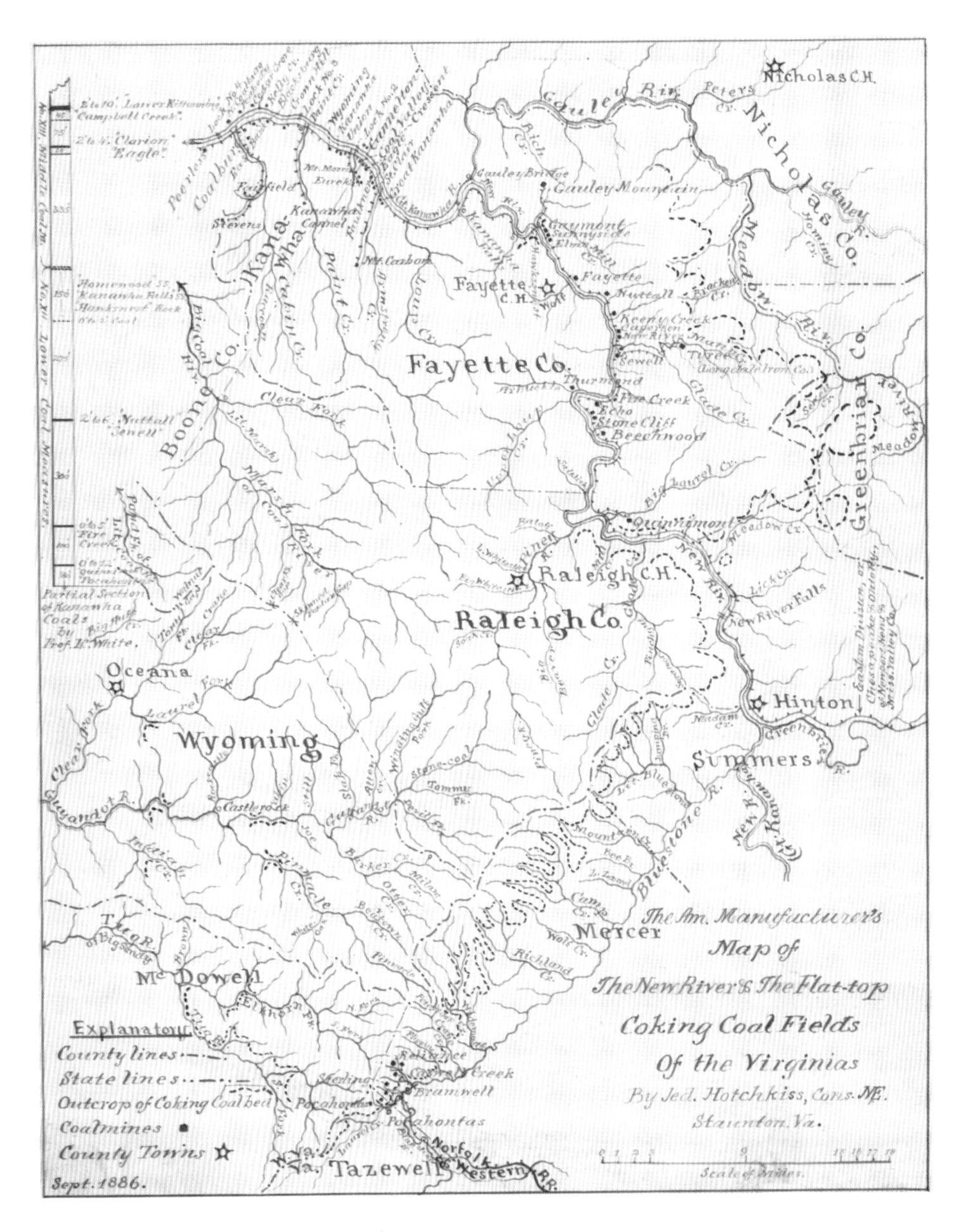

The Am. Manufacturer's map of the New River & the Flat-top coking coal fields of the Virginias, by Jed. Hotchkiss, Cons. M.E., Staunton, Va., Sept. 1886.

Iron was being produced by the use of coke instead of charcoal and shipped to the Atlantic port of Newport News, as indicated by the notation at right of the "Eastern division of the Chesapeake and Ohio RR of Newport News and Miss. Valley Co." Locks on the James River and Kanawha Canal are labeled. The Great Kanawha River joins the New River, and the railroad parallels the canal to Richmond.

Postwar Recovery and Growth

Plans for reconstruction of the Confederate states were under way before General Lee's surrender. Though President Lincoln had proposed that when 10 percent of the voters in each state had sworn allegiance to the United States, the state could return to the union, Republicans in Congress wanted to punish the South for the costly war. Their position was that since the states had broken away, they must meet stiff standards in order to be readmitted. They proposed the Wade-Davis Bill, which stripped the right to vote and hold office from all white males who had "taken up arms" against the union. This meant essentially everybody except those whites too old for the army.

Governor Francis Harrison Pierpont of the newly formed state of West Virginia proposed that he be in charge of returning Virginia to the union, but Lincoln said that the western counties' separation had been a "war measure" and was temporary. (Lincoln essentially felt that the union was an unbreakable entity and that Southern states had never legally been free of it—and he felt that if the union was unbreakable, so was a state.) The fate of the former Confederacy hadn't been settled when Lincoln was assassinated only a week after Lee's surrender at Appomattox. The new president, Andrew Johnson, was unpopular with Congress and lacked Lincoln's ability to persuade and compromise.

President Johnson's plan for reconstructing the South was a compromise between Lincoln's plan and the Wade-Davis Bill. He appointed provisional governors for each of the states—Pierpont was named Virginia's governor—and elections were held for other officials. Johnson disliked the large planters, whom he thought were responsible for the war, but he believed in states' rights and private property. He issued thirteen thousand pardons to individual former Confederates and later a blanket pardon, excluding only those who had property worth more than $20,000 or who had held high office.

Some Radical Republicans wanted Confederate president Jefferson Davis, other high officials, and high-ranking military officers hanged for treason, but they realized that if Davis were hanged, he would become a martyr and the South would be even more embittered than it already was. Davis spent two years as a prisoner in Fort Monroe at Hampton, Virginia, part of the time in leg irons. He was eventually freed and allowed to return to Mississippi.

Virginia was devastated by the war. An entire

generation of her young men had been killed. The capital city, Richmond, lay in ruins. Railroads had been torn up—sometimes by the Confederates themselves, so that they wouldn't fall into the hands of the invading Union armies. Houses, barns, and businesses had been burned; fields had grown up in weeds or brush; and animals had been stolen or slaughtered. Confederate money was worthless, as were bonds issued by the Confederacy. One of the major assets of Virginia planters—slaves—had been freed, and the owners would not be compensated. And with the formation of West Virginia, Virginia had lost three hundred thousand of its citizens and a third of its territory and taxable property.

The situation was too much for some Virginians. Edmund Ruffin, an innovator in agriculture before the war, saw no future for himself in Reconstruction-era Virginia and committed suicide. General Lee, however, showed gallantry in defeat. His family's estate, Arlington, was seized by the Union, and Lee was forbidden to ever set foot on the estate again. Lee was invited to emigrate to Europe, Mexico, or Brazil, and while some former Confederates did so, Lee chose to remain in Virginia. Lee's wife was ill, the general's health had deteriorated, and the family was destitute. A friend offered them a modest home, which Lee accepted until he became the president of Washington College in Lexington, Virginia—now Washington & Lee University.

Of those who left the state, Matthew Fontaine Maury, the mathematician and oceanographer known as the Pathfinder of the Seas, first went to Mexico, then to England, and then returned to Virginia to teach at Virginia Military Institute and urge immigrants to come to Virginia. His work in mapping the ocean was crucial for laying the transatlantic cable.

Former slaves wandered, confused about their condition and lacking skills, education, and assets. The Freedmen's Bureau was established to help them. Food was distributed to thousands in Virginia, and teachers came from Northern states to teach them in makeshift schools, as there were no public schools for either race. Young men drifted into towns, leaving behind women and children and the aged to fend for themselves, often on the farms where they had been enslaved, dependent on their former masters. Blacks who had served in the Union army retained their weapons, and others were armed by Unionists. Virginians feared an insurrection, and the state legislature passed laws against vagrancy and violence. Radical Republicans in Congress accused the South of planning a return to slavery, though the Thirteenth Amendment to the Constitution had ended slavery in the United States. Not mentioned was that Virginia had also passed laws requiring witnesses to explain contracts to illiterate blacks so that they would not be exploited.

Congress next proposed the Fourteenth Amendment, making the former slaves citizens. In order to be readmitted to the union, the former Confederate states had to ratify this amendment. President Johnson didn't push ratification, as the Southern states had had no part in writing the amendment and most of their citizens were unable to vote. Radical Republicans accused him of sympathy to the South since he had owned slaves himself. Congress brought impeachment charges against President Johnson, but his removal failed by one vote. For the remainder of his term in office, he vetoed punitive legislation against the South, only to have his vetoes overridden.

On March 2, 1867, statehood was stripped from Virginia, and it became Military District 1, under the command of Gen. John M. Schofield. Unionists attempted to disenfranchise as many Virginians as possible, and even minor officials were subjected to questioning on how much they had aided the Confederacy. Had they provided food or clothing to any rebels? If they'd been soldiers, were they conscripted or had they volunteered? The effect was to keep 95 percent of Virginia's white male residents from voting. Governor Pierpont, who at first supported Unionist rules, saw that this wouldn't leave enough competent people to administer the state. When he presented his objections to Congress, he was removed as governor.

In April 1867, Federal judge John Underwood, a radical who accused Virginians of burnings and assassinations, and of deliberately spreading smallpox and yellow fever, called a constitutional convention for the state in Richmond. Presiding as delegates were mainly freed blacks, Unionist whites newly arrived in Virginia, and six from foreign countries. General Schofield protested the situation to his former colleague, Gen. Ulysses Grant, now U.S. president. Because of the efforts of Schofield and a group of conservative Virginians, the worst provisions of the Underwood Constitution were defeated. In October 1869 the Virginia General Assembly approved it and ratified the Fourteenth and Fifteenth Amendments to the U.S. Constitution, though they felt that the Fifteenth, which guaranteed the freed slaves the right to vote, was unfair to the South, since only five of the other states in the union allowed all blacks to vote. On January 16, 1870, Virginia was readmitted to the union.

Virginia, deep in debt, began to pick up the pieces. Some in the legislature, called the Funders, said that for the state's honor, Virginia must repay the debts incurred in building roads, railroads, and canals before the war. Interest on the bonds had been accruing throughout the war, for a total of $45 million. Full payment would take the entire state budget, leaving nothing for badly needed schools, hospitals, and other facilities, or for operating the state government. Since the debt had been taken on while the western counties were still a part of the state, the Virginia legislature thought that West Virginia should pay a third of the debt. It refused to do so until forced by the Supreme Court—fifty years later. A group called the Readjusters succeeded in having the debt amount reduced, but new bonds were issued with 6 percent interest, a rate higher than the original. To help pay off the debt and keep the state functioning, Virginia sold off its railroads, which was to prove a bad move.

Virginia suffered along with the rest of the country from periodic financial panics, or recessions. Prices for farm produce plummeted. Money was scarce, and taxes were mainly assessed on property. Some planter families abandoned their land and moved to cities, where their sons became lawyers or businessmen. Large holdings were broken up so that the average farm could be worked by a single family. Those who owned no land, including former slaves, often became sharecroppers, working on someone else's farm. The landowner got a percentage of the crop and might furnish food and supplies until the harvest. The workers were poor but free and could move to another farm for better opportunities. Their children could go to school, for at least a few months a year, since Virginia had at last set up free public schools for both races.

Huge changes had taken place in Virginia since 1860, and more changes were yet to come.

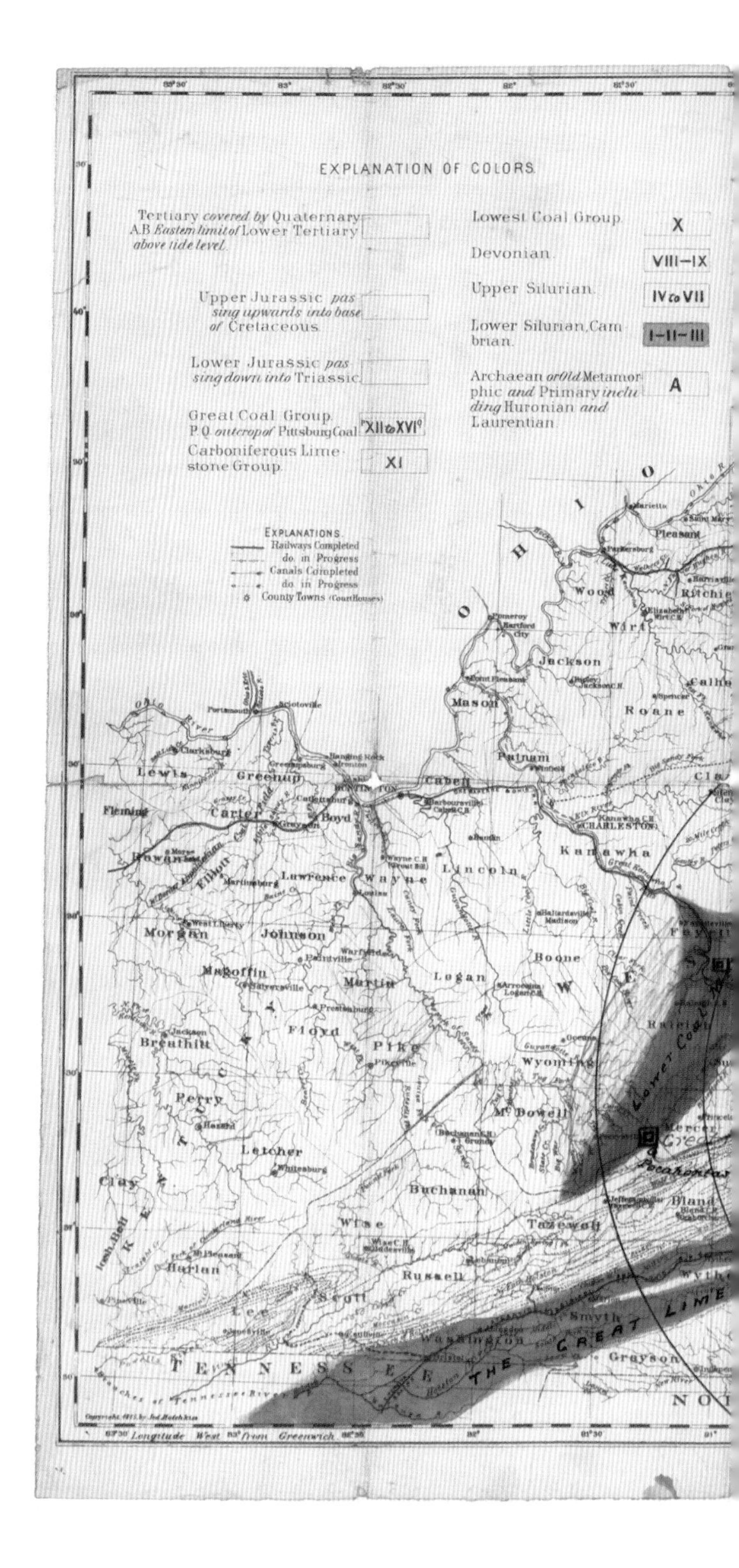

Hotchkiss' geological map of Virginia and West Virginia, the geology by Prof. William B. Rogers, chiefly from the Virginia State survey, 1835–41, with later observations in some parts.

Geological map of Virginia and West Virginia, 1875, showing the "great Limestone Valley" of Virginia and the location of two coal mines. Iron ore was abundant and easily mined in the area. Using coke made from coal, limestone, and ore, huge quantities of pig iron were produced in Virginia, especially at Lucy Selina Furnace in Alleghany County.

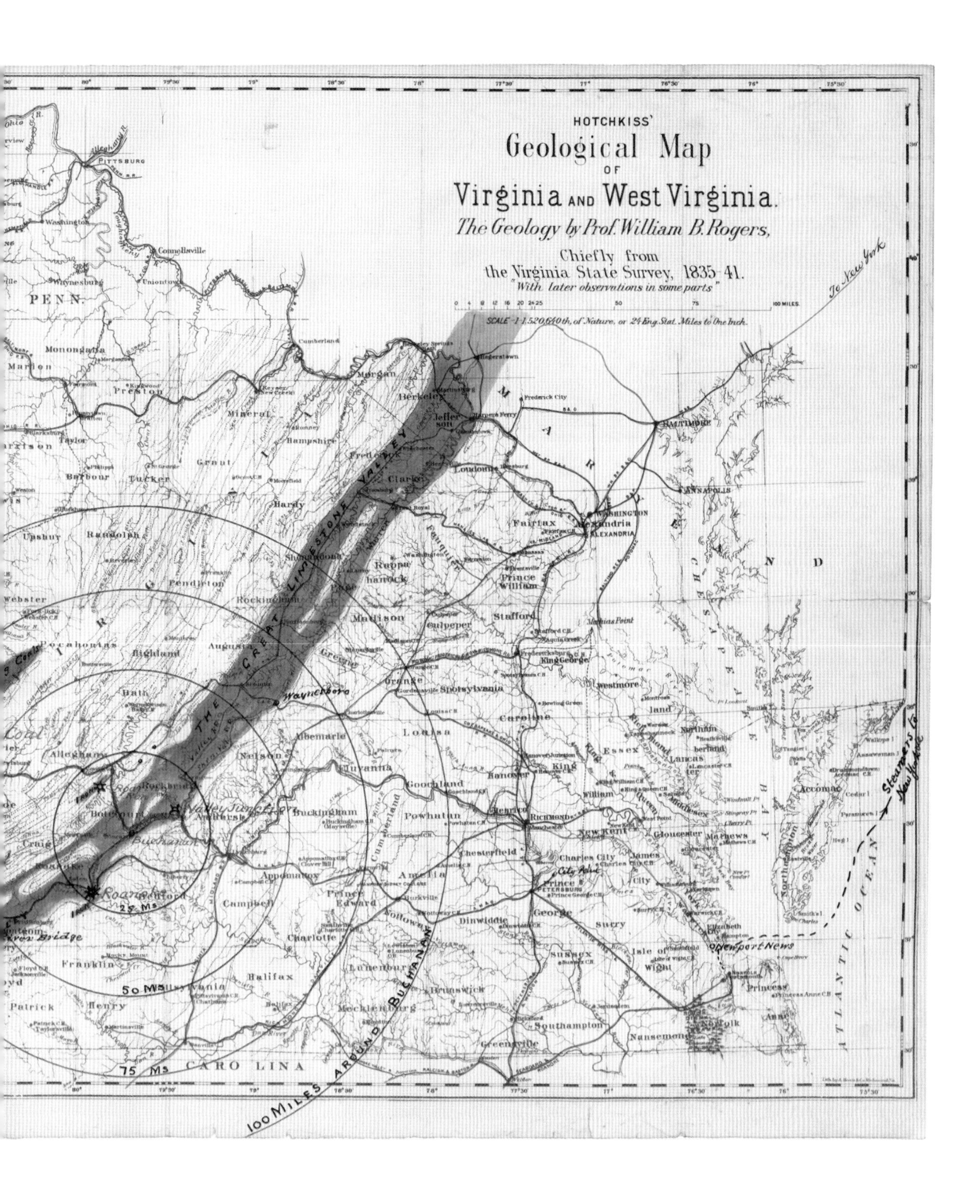

HOTCHKISS'
Geological Map
OF
Virginia AND West Virginia.
The Geology by Prof. William B. Rogers,
Chiefly from
the Virginia State Survey, 1835-41.
"With later observations in some parts"
SCALE - 1:1,520,640th, of Nature, or 24 Eng. Stat. Miles to One Inch.
To New York
Steamers to New York
THE GREAT LIMESTONE VALLEY
100 MILES AROUND BUCHANAN
75 Ms
50 Ms
25 Ms
Waynesboro
Newport News
ATLANTIC OCEAN
CHESAPEAKE BAY
PENN.
CAROLINA
MARYLAND

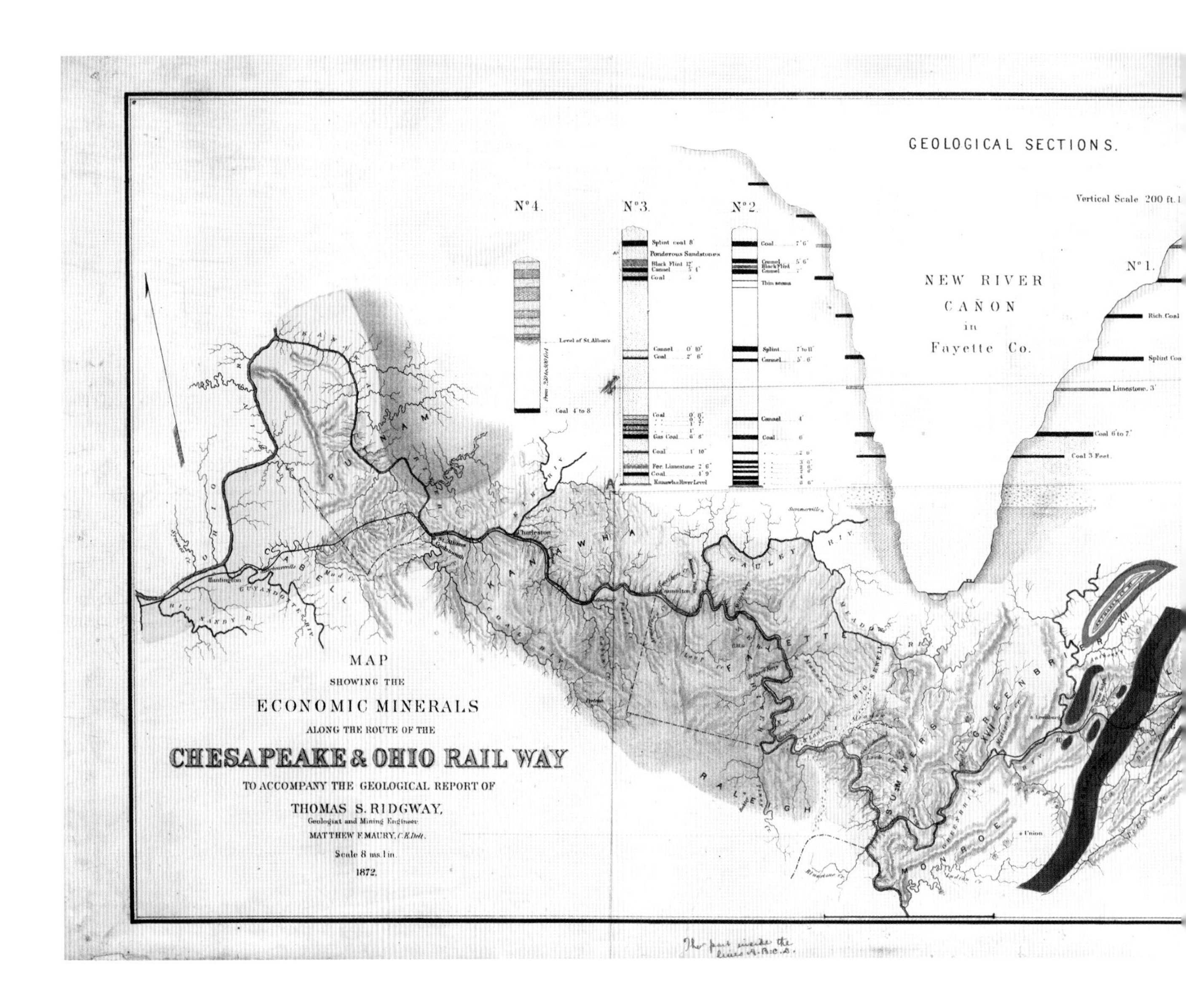
GEOLOGICAL SECTIONS.
Vertical Scale 200 ft.
N°4.
N°3.
N°2.
N°1.
NEW RIVER
CAÑON
in
Fayette Co.
MAP
SHOWING THE
ECONOMIC MINERALS
ALONG THE ROUTE OF THE
CHESAPEAKE & OHIO RAIL WAY
TO ACCOMPANY THE GEOLOGICAL REPORT OF
THOMAS S. RIDGWAY,
Geologist and Mining Engineer.
MATTHEW F. MAURY, C.E. Delt.
Scale 8 ms. 1 in.
1872.
PUTNAM
KANAWHA
CABELL
FAYETTE
RALEIGH
SUMMERS
GREENBRIER
MONROE

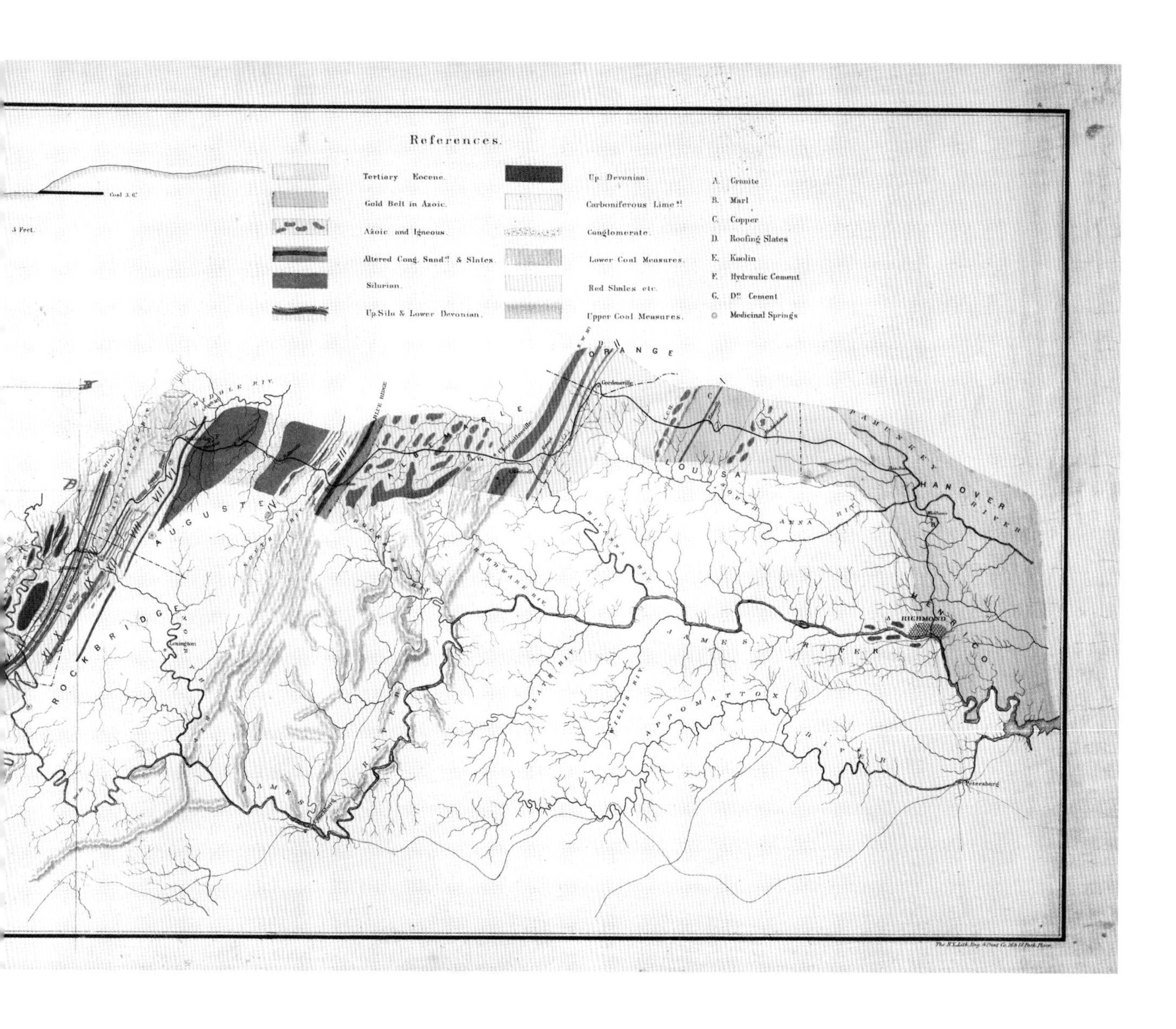

Map showing the economic minerals along the route of the Chesapeake & Ohio Rail Way to accompany the geological report of Thomas S. Ridgway.

This 1872 map shows economic minerals found along the route of the Chesapeake and Ohio Railway from Richmond, Virginia, to the Ohio River. The heavy black line is the border between Virginia and West Virginia. Coal was shipped from West Virginia to smelters in Virginia, and the finished product was shipped both east and west. Map by Matthew Fontaine Maury, with geological information from Thomas Ridgway.

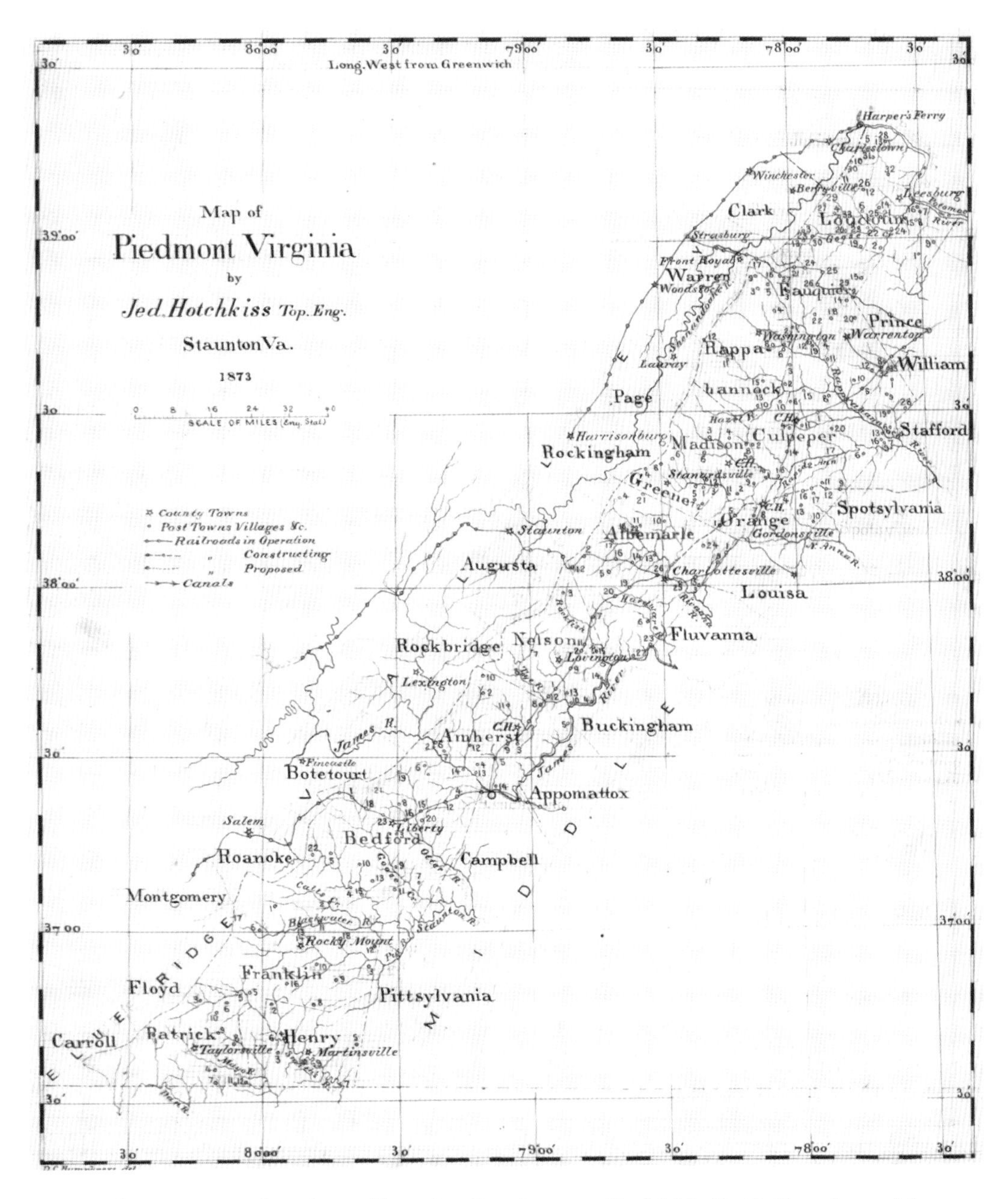

Map of Piedmont Virginia, by Jed. Hotchkiss, Top. Eng. Staunton, Va. 1873; D. C. Humphreys del.

A map of Piedmont Virginia from southwest Loudoun County to Carroll County, Virginia. Elevation is not shown by shading, but by altitude numbers written in various locations. Towns, post offices, and villages are labeled, and the map key has symbols for railroads in operation, railroads being constructed, and railroads proposed, an indication that this was an era of railroad building in Virginia.

Map of middle Virginia, by Jed. Hotchkiss, Top. Eng., Staunton, Va., 1873.

A map of middle Virginia, from northwest Fairfax County to southwest Pittsylvania and southeast to Greensville County, Virginia, by Jedediah Hotchkiss (1828–1899), geologist and renowned mapmaker. By 1873, there were at least three railroads serving Richmond and two serving Petersburg, and the James River and Kahawha Canal was still in operation. Alexandria is listed as a county rather than a city.

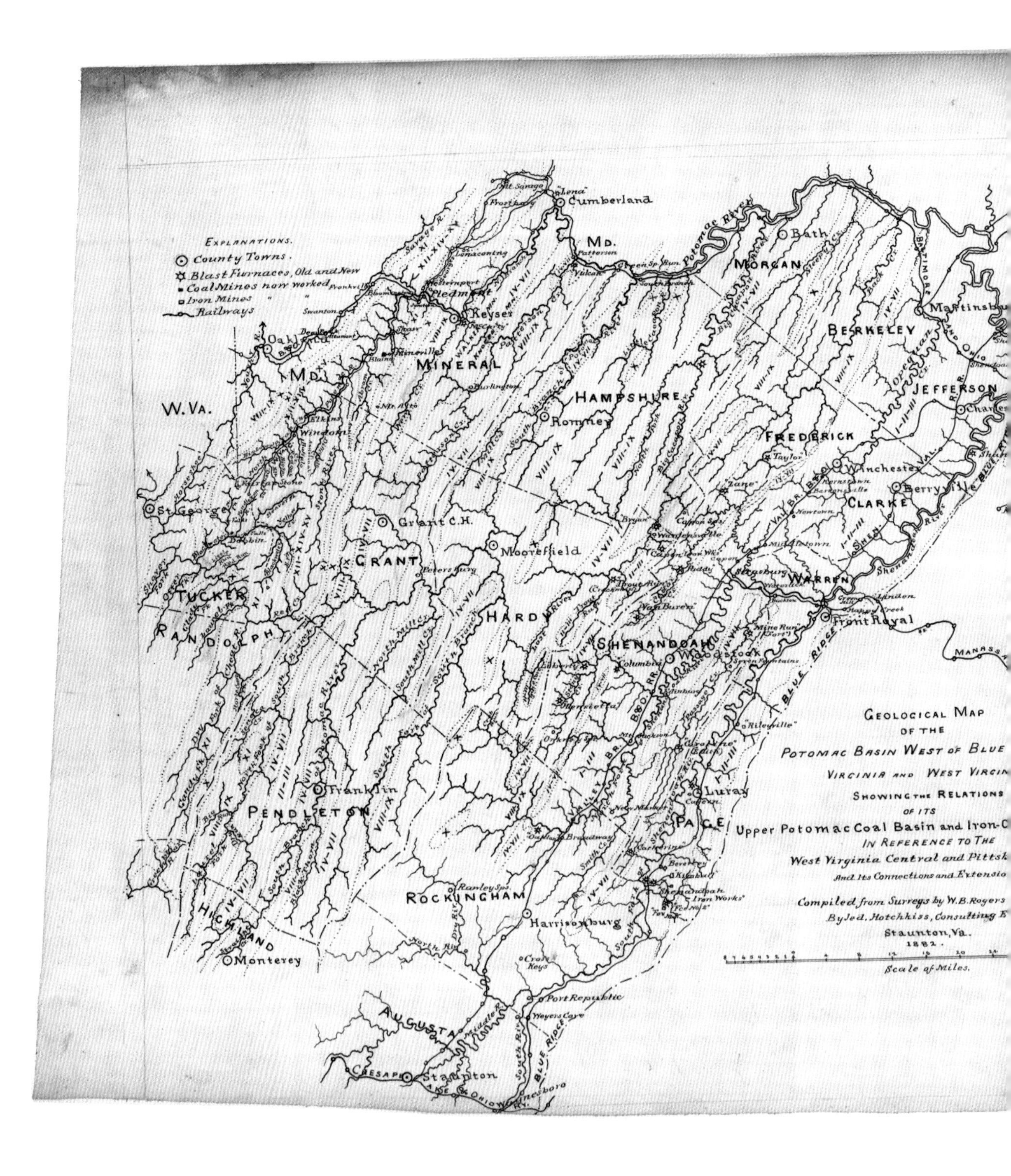
Explanations.
County Towns.
Blast Furnaces, Old and New
Coal Mines now worked
Iron Mines " "
Railways
Cumberland
Md.
Potomac River
Bath
Morgan
Piedmont
Keyser
Oakland
Md.
Mineral
Berkeley
Martinsburg
Jefferson
W. Va.
Hampshire
Romney
Frederick
Winchester
Berryville
Clarke
St. George
Grant C.H.
Moorefield
Grant
Tucker
Randolph
Hardy
Warren
Front Royal
Shenandoah
Woodstock
Luray
Page
Franklin
Pendleton
Rockingham
Harrisonburg
Highland
Monterey
Port Republic
Augusta
Staunton
Blue Ridge
Geological Map
of the
Potomac Basin West of Blue
Virginia and West Virgin
Showing the Relations
of its
Upper Potomac Coal Basin and Iron-C
In Reference to The
West Virginia Central and Pittsb
And Its Connections and Extensio
Compiled from Surveys by W.B. Rogers
By Jed. Hotchkiss, Consulting E
Staunton, Va.
1882.
Scale of Miles.

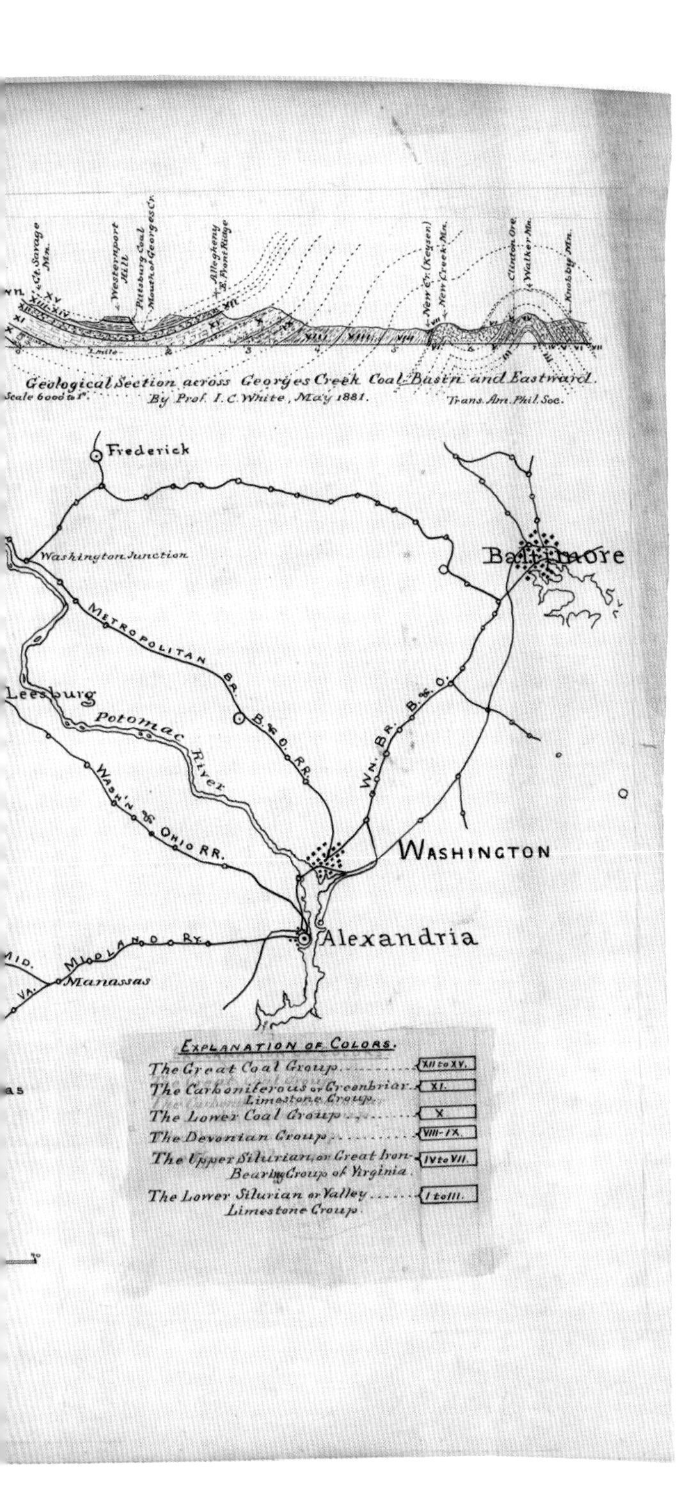

Geological map of the Potomac basin west of Blue Ridge, Virginia and West Virginia: showing the relations of its upper Potomac coal basin and the iron-ore bearing areas in reference to the West Virginia Central and Pittsburg [sic] R.R., and its connections and extensions, compiled from surveys by W. B. Rogers and others by Jed. Hotchkiss, Consulting Eng., etc.

Of particular interest in this 1882 map are the various railroads that converge on Washington, D.C., and Alexandria, Virginia: the Valley Branch, Manassas Branch, and Metropolitan Branch of the Baltimore & Ohio, as well as the Virginia Midland and the Washington & Ohio.

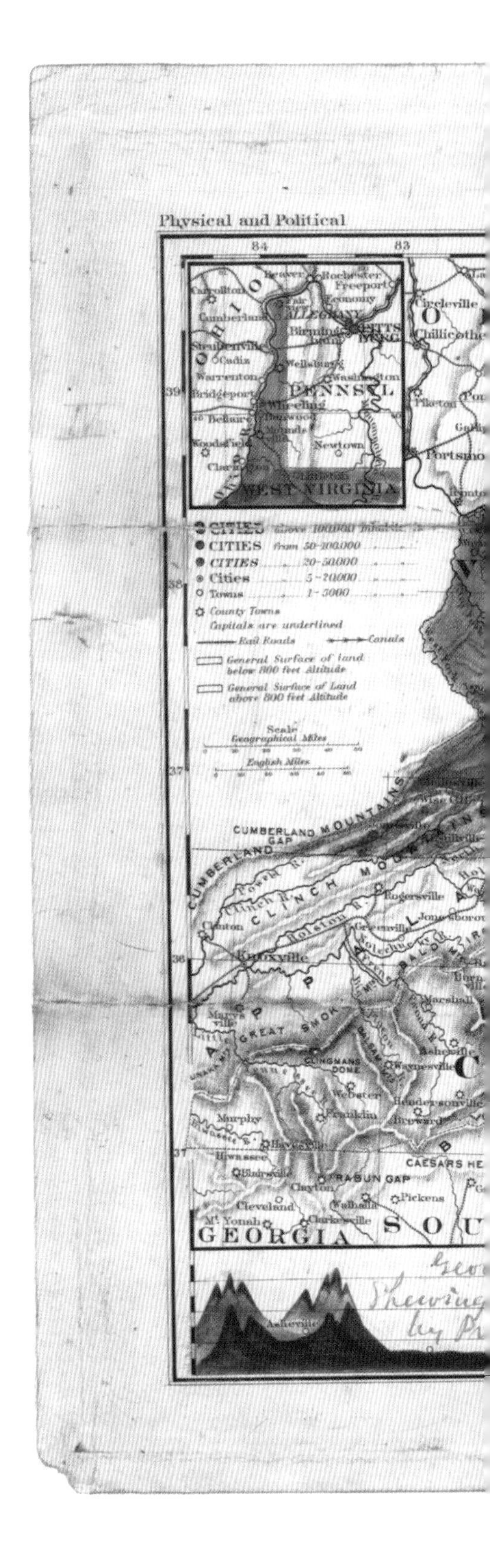

Geological map of Virginia & West Virginia showing their chief geological sub-divisions, by Prof. William B. Rogers on basis of the physical & pol. map of A. Guyot.

The creator of this map, ca. 1870, William Barton Rogers, was Professor of Natural Philosophy (geology) at William & Mary and at the University of Virginia, and founder of MIT. By use of color, Rogers indicates the geologic period in which the various areas of Virginia were formed and what minerals are likely to lie beneath the surface. For instance, all of West Virginia is labeled "carboniferous," as containing coal and iron ore, and a similar indication is made for an area just west of Richmond.

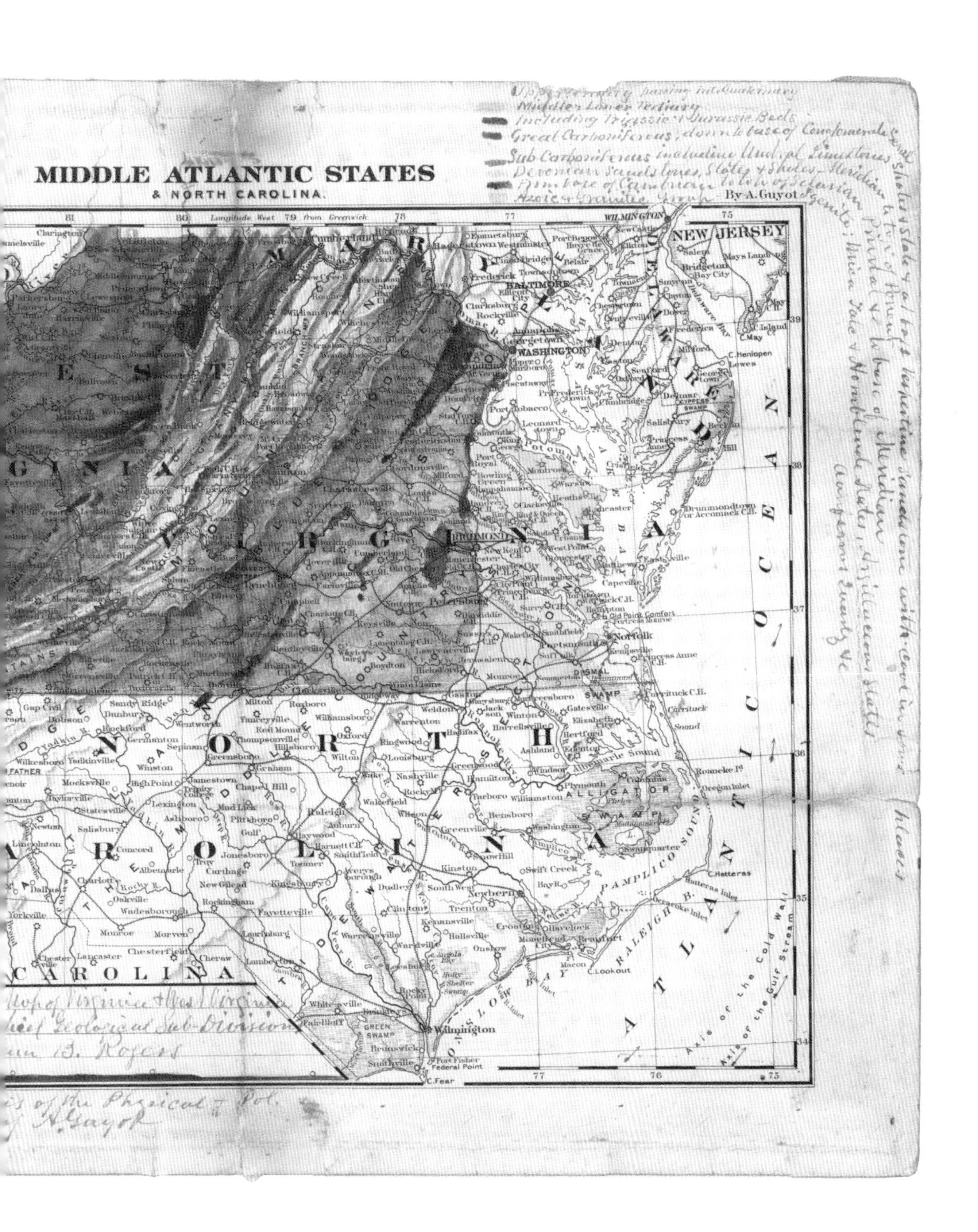
MIDDLE ATLANTIC STATES
& NORTH CAROLINA.
By A. Guyot
Longitude West 79 from Greenwich
NEW JERSEY
WASHINGTON
BALTIMORE
RICHMOND
Petersburg
Norfolk
Raleigh
Wilmington
WILMINGTON
DISMAL SWAMP
ALLIGATOR SWAMP
GREEN SWAMP
PAMPLICO SOUND
RALEIGH B.
ONSLOW BAY
C. Hatteras
C. Lookout
C. Fear
C. Henlopen
C. May
Old Point Comfort
Axis of the Cold Wall
Axis of the Gulf Stream
ATLANTIC OCEAN
CAROLINA
Albemarle Sound
Currituck Sound

McDowell County, W. Va.
Great Flat-top Mtn.
Peeled Chestnuts Br.
West Fk.
East Fork
"Reliance" Colliery
Wm. Booth & Co.
Stevenson, Mullin, & Co.
Caswell Creek Coal & Coke Co.
Caswell Fork
Mercer County, W. Va.
Simmons Cr.
Bluestone River.
Simmons
"Sterling" Colliery John Cooper & Co.
W. Va.
Va.
West Fork
N. Fork
Outcrop of "Pocahontas" Coal Bed.
Mills
Bramwell
Cooper
Map of Part of The Great Flat-top Coal-field of Va. & W. Va. Showing Location of Pocahontas & Bluestone Collieries May, 1886.
Peeled Chestnuts Gap
Tazewell County, Va.
Coal Br.
East Mine
West Mine
Bluestone Junction
New R. Br. o Norfolk & W'n.
Bluestone River.
0 1/4 1/2 1 Mile
Scale
Pocahontas
Laurel Creek
Eng. Office of Jed. Hotchkiss, Staunton, Va.

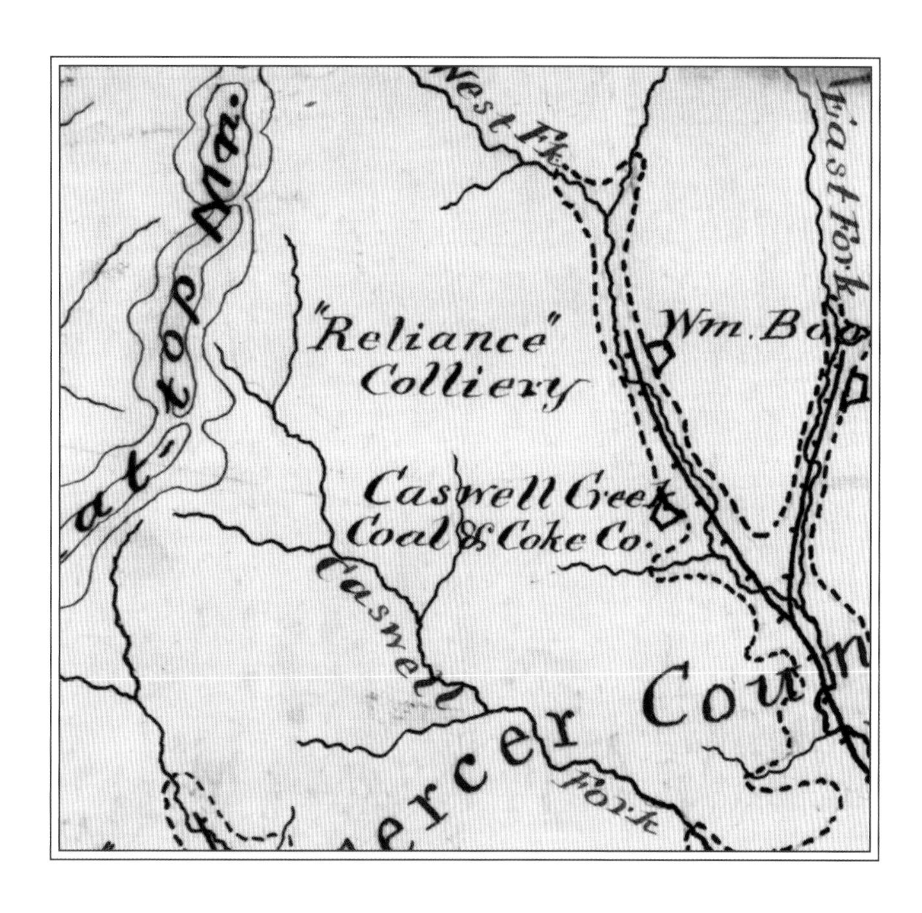

Map of part of the great Flat-top coal-field of Va. & W. Va. showing location of Pocahontas & Bluestone collieries, May 1886, Eng. Office of Jed. Hotchkiss.

This map (at left), of the great flat-top coalfield of Virginia and West Virginia, shows the location of the huge Pocahontas colliery in Tazewell County, Virginia, and the Bluestone and several other collieries in Mercer County, West Virginia. At bottom right is the Norfolk & Western Railway, which, with its several branch lines, carried much of the coal from the mountains to seagoing vessels.

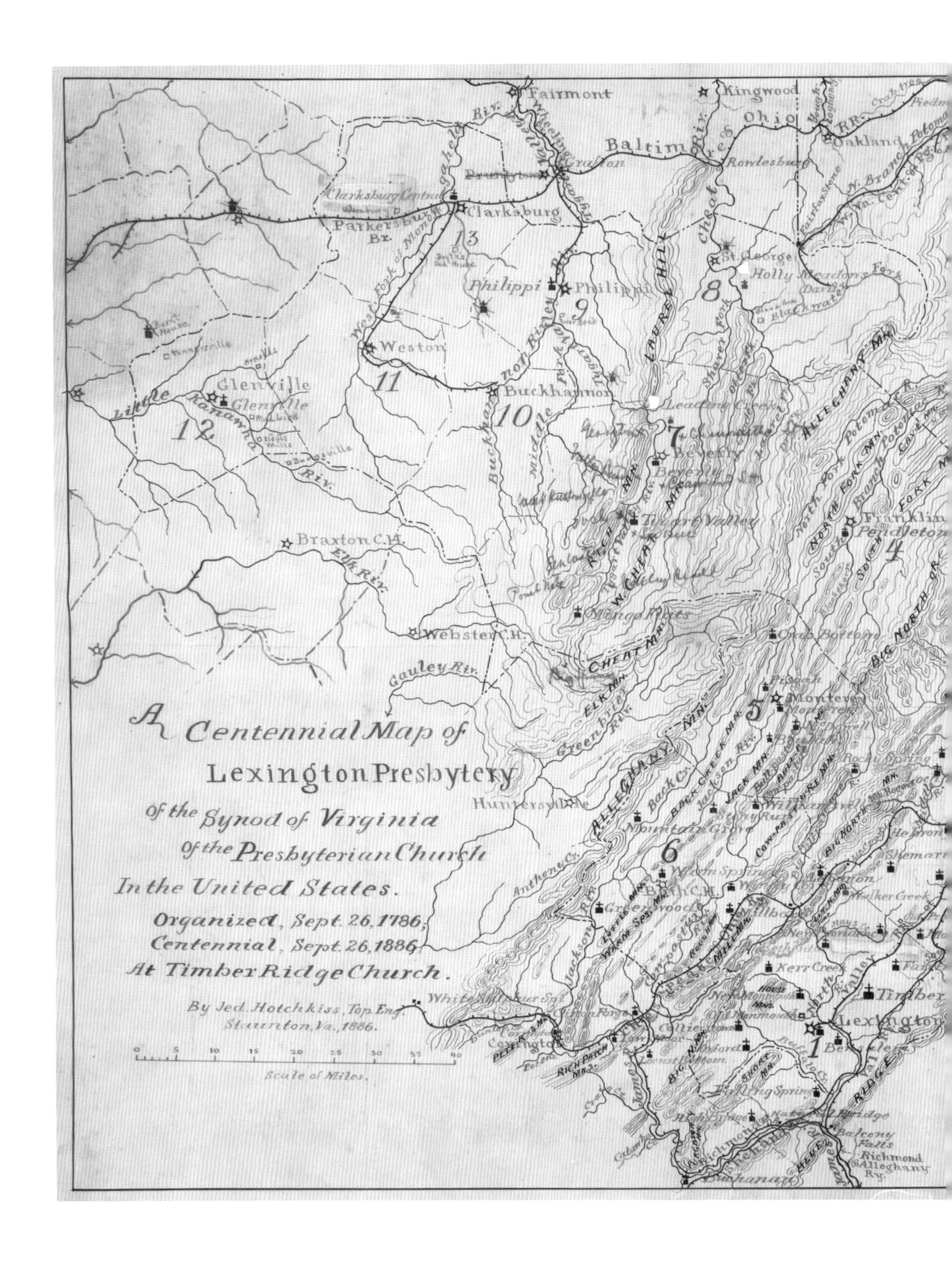
A Centennial Map of
Lexington Presbytery
Of the Synod of Virginia
Of the Presbyterian Church
In the United States.
Organized, Sept. 26, 1786;
Centennial, Sept. 26, 1886;
At Timber Ridge Church.
By Jed. Hotchkiss, Top. Eng.
Staunton, Va., 1886.
Scale of Miles.
Fairmont
Kingwood
Clarksburg
Grafton
Rowlesburg
Oakland
Philippi
St. George
Weston
Buckhannon
Glenville
Beverly
Franklin
Braxton C.H.
Webster C.H.
Huntersville
Lexington
Buchanan
Covington
Balcony Falls
Richmond & Alleghany Ry.

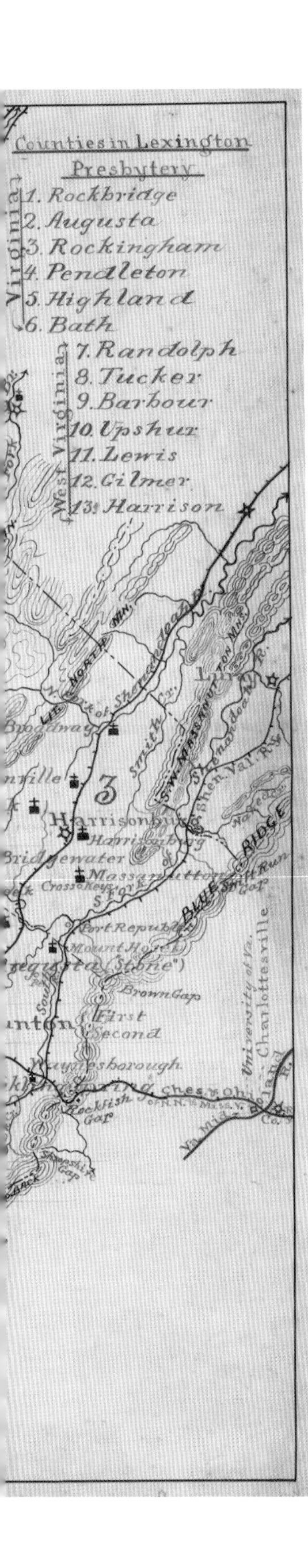

A centennial map of the Lexington Presbytery of the Synod of Virginia of the Presbyterian Church in the United States, organized Sept. 26, 1786, centennial Sept. 26, 1886, at Timber Ridge Church, by Jed. Hotchkiss, Top. Eng.

The numerous churches of the Lexington Presbytery are named and designated by a tiny church symbol on this centennial map. Most settlers of the mountain and valley region were Protestant, and many, being Scotch-Irish, were Presbyterian. The Virginia counties included are Rockbridge, Augusta, Rockingham, Highland, Bath, and—strangely—Pendleton, though there is no Pendleton County in Virginia. Also included are the West Virginia counties of Randolph, Tucker, Barbour, Upshur, Lewis, Gilmer, and Harrison, all named for distinguished antebellum Virginians.

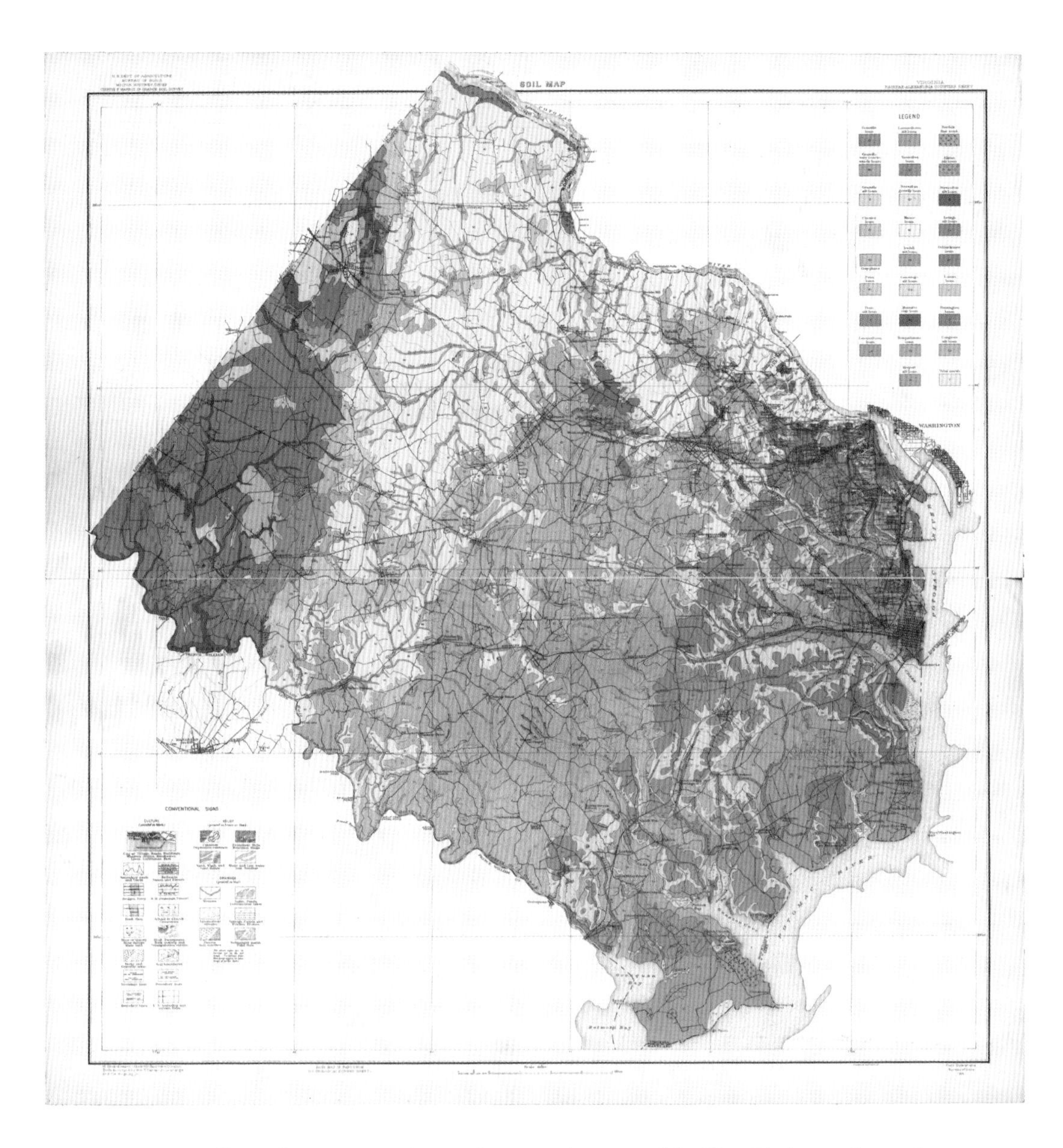

Virginia, Fairfax-Alexandria counties sheet, soil map; soils surveyed by Wm. T. Carter Jr., in charge, and C. K. Yingling Jr. Map of Fairfax and Arlington Counties and the City of Alexandria, prepared in 1915 by William T. Carter for the U.S. Department of Agriculture, Bureau of Soils.

Despite its proximity to the nation's capital, this part of the state, like most of Virginia, was still largely rural and dependent on agriculture. The explosive growth of the late twentieth century had not occurred. Mount Vernon and Gunston Hall, former homes of George Washington and George Mason, respectively, are at the bottom right. The Episcopal Seminary had been built in Alexandria, but Fort Belvoir and the Pentagon had not been built.

Entering the Twentieth Century

After Virginia was readmitted to the union in 1877, the state still faced a long struggle to regain her prosperity and stature. Virginia was the richest of the former Confederate states, but income per resident was still $100 less than that of Kansas, the poorest of the non-Confederate states.

Education improved for both races, though for a good while the improvement was mainly at the college level. Rural elementary and high schools were badly equipped, the school year was often short, and teachers were underpaid, but there was a push to educate all children. For those who graduated from high school, state-supported colleges were available. Virginia Polytechnic Institute (now Virginia Polytechnic Institute and State University) had been established in 1872 as part of the Federal Land Grant Program, setting up schools for mechanical arts and agriculture in each state. The College of William & Mary in Williamsburg was revived after a seven-year lapse, and throughout the commonwealth junior colleges and "normal schools" educated generations of Virginia teachers.

Hampton Institute was established by the Freedmen's Bureau as a school for blacks. Its most famous graduate, Booker T. Washington, was a former slave from Franklin County, Virginia. After working his way through Hampton Institute, he went to Tuskegee Institute in Alabama and made it the outstanding institution it is. He became a spokesman for blacks, urging them to learn a trade and become self-supporting. He was the first black inducted into the American Hall of Fame and was a consultant to Presidents Theodore Roosevelt and William H. Taft. His autobiography, *Up from Slavery,* is still popular.

Also important in black education was John Mercer Langston, who was born in Virginia but educated in Ohio after his parents died. He was the dean and president of Howard University and U.S. minister to Haiti. In 1885 he was chosen president of Virginia Normal and Collegiate Institute in Petersburg, Virginia's first state-supported college for blacks. He upgraded the school, but as a Republican, he was removed by the Democratic legislature when his term ended.

Virginia developed commercially at this time, mainly because of railroads. Colis P. Huntington bought the C&O Railroad and extended it from Richmond to Newport News. Here he built coal and grain storage, and Newport News Shipbuilding and Drydock, one of the largest shipyards in America. In 1900, Huntington's shipyard was valued at $15 million.

When a rich vein of coal was discovered in southwestern Virginia, investors quickly built railroads to take the coal to Norfolk. The Norfolk and Western Railroad and the Virginian ran along similar routes from the coalfields to the coast. Peanuts, cotton, and tobacco also arrived by rail, and timber and other produce came from North Carolina on the Dismal Swamp Canal. It all went from the port of Norfolk to the rest of the world in twenty-six steamship lines. In addition, vessels plied the Chesapeake Bay, taking passengers, garden produce, and seafood to northern cities.

Richmond, the terminus of both north–south and east–west railroads, had over 1,200 manufacturing plants by 1900 and had the world's first electric street trolleys. Nearby Petersburg became an inland port, with ample electricity from hydroelectric dams.

Because of the railroads, the settlement of Big Lick, which in 1881 had four hundred residents, became the City of Roanoke. By 1892 its population topped twenty-five thousand, and it doubled again within the next decade. Throughout western Virginia speculators touted the bright prospects of proposed cities, and one such site was Big Stone Gap. Both coal and iron ore were discovered in the area, and by 1890 a railroad reached the town. British and northern investors flocked to the town, plans were afoot for a huge hotel, and the Duke and Duchess of Marlborough visited (but bought lots elsewhere in Virginia). Alas, the iron ore was of poor quality, and the financial panic of 1893 ended the dreams of developers, in Big Stone Gap and in other possible towns.

Manufacturing grew, with areas specializing: Danville produced textiles; Lynchburg, tobacco products; Martinsville and Bassett, furniture; and Hopewell, chemicals. Norfolk had shipping, but also a robust fertilizer industry. James Bonsack invented a cigarette rolling machine, which gave a boost to the tobacco industry.

A Virginia native, Thomas Fortune Ryan, a tenant farmer's son from Nelson County, made his fortune elsewhere. He bought a seat on the New York Stock Exchange at twenty-three and owned street railways, city lighting companies, coal mines, insurance companies, and even a Congolese diamond mine. When he retired in 1908 as director of more than thirty companies, he returned to Nelson County and his home, Oak Ridge, where he had a horse-racing track, railway station, a church, and livestock operations on thousands of acres. He entertained presidents, business leaders, and foreign heads of state, who all arrived in Ryan's private railroad car.

Wealthy northerners were returning to Virginia's spas, such as Hot Springs, or to the newly built Chamberlin Hotel at Hampton, where they met Virginia's belles. One symbol that North–South enmity was ending was the fashionable marriage between painter Charles Dana Gibson and Irene Langhorne of Albemarle County. Her sister Nancy married Lord Astor and became the first woman elected to the British House of Commons.

But despite urban prosperity, by 1900 Virginia was still 85 percent rural, and for farmers there had been no boom times. Prices of produce stayed low, and they were at the mercy of railroads and merchants of fertilizer and farm equipment.

A new state constitution in 1902 established a state corporation commission to regulate the railroads and other corporations. It also established an Alcoholic Beverage Control board, which mandated that alcohol could only be sold in state-run stores and only in full, sealed containers—no liquor by the drink. The same constitution gave the largely Democratic legislature the power to appoint judges and election officials, and voters were required to pay a poll tax of $1.50 per year for three years prior to voting, and to pass a literacy test. Both black and white voting rates dropped sharply. In addition, blacks faced Jim Crow laws that mandated separate waiting rooms, train cars, seating in theaters, and even restrooms and water fountains.

Some exceptional blacks succeeded despite the system, such as Maggie Walker of Richmond, who became wealthy when she established a "penny" bank and an insurance business, but for most blacks, the opening of the twentieth century was a dismal time. Many moved north to cities such as New York, Detroit, and Chicago.

On the national scene Virginians achieved prominence, though they often failed financially. Dr. Walter Reed worked with the builders of the Panama Canal to find the cause and cure for yellow fever, and he saved the lives of countless people. Dr. Crawford Long used anesthesia on his patients before Dr. W. T. G. Morton, who gets credit for the procedure, and Mahlon Loomis set up a wireless system in the Lynchburg area before Marconi was born, but he could not get financial support for his project.

In the literary world, Thomas Nelson Page wrote best-selling novels, Mary Johnston wrote historical romances based in Virginia, and Ellen Glasgow's novel, *The Descendant,* about Virginia families, was published in 1897. She would later win the Pulitzer Prize for literature. John Fox Jr. wrote the popular *Trail of the Lonesome Pine,* which brought attention to southwest Virginia and the conflict between mountain settlers and businessmen.

Virginians began to take pride in their history. Lee was revered, statues to various Civil War generals were erected along Monument Avenue in Richmond, and Stonewall Jackson's widow and daughter Julia were welcomed throughout the commonwealth and given free railroad passes and financial support. In 1888, two women, Mary Galt and Mrs. Charles Coleman, bought the Powder Horn in Williamsburg to save it from destruction. The Association for the Preservation of Virginia Antiquities was formed and restored the ruined church at Jamestown. In 1900 the entirety of Jamestown Island was set aside for preservation.

The celebration of the centennial of the surrender of the British at Yorktown in 1881 brought many visitors from the North to Virginia, for Washington's victory was the whole nation's victory. A decade later the International Naval Rendezvous, celebrating the four-hundredth anniversary of Columbus's voyage to America, brought ships and thousands of visitors to the Hampton Roads harbors.

With the Spanish-American War in 1898, America became a world power, and Virginia was fully a part of the nation. Virginian Fitzhugh Lee, nephew of Gen. Robert E. Lee, was American consul general in Cuba, and Virginia's young men joined the forces being embarked through Norfolk to fight in the brief war.

A decade later, in 1907, people from all over America and half a dozen foreign countries joined Virginia in celebrating the three-hundredth anniversary of the settlement at Jamestown, where America began. Woodrow Wilson, born in a parsonage in Staunton, Virginia, and educated at the University of Virginia, would become U.S. president five years later.

The twentieth century, with its wars, its inventions, and its amazing changes, had begun, and Virginia was an active participant.

Perspective map of the city of Roanoke, Va. 1891. American Publishing Co. (Milwaukee, Wis.).

This 1891 map of Roanoke shows the city's rapid growth in the late nineteenth century, mostly because of railroads. Roanoke then had fourteen churches, seven hotels, four newspapers, an academy of music, an opera house, and three main railroads, as well as a narrow-gauge railroad and two dummy lines, and fifty-four manufacturers. These included flour mills, brewers, bridge works, a tobacco factory, several foundries, and factories to make paper bags, mattresses, bricks, tiles, and spikes.

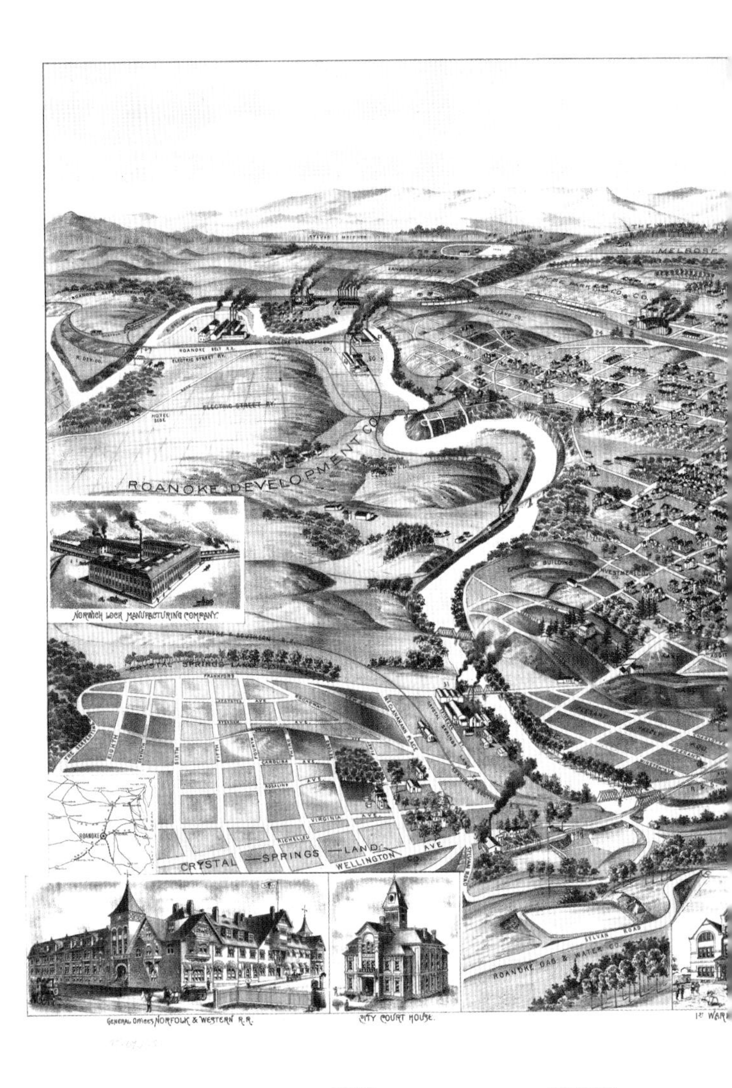

PERSPECTIVE MAP OF THE CITY OF

OANOKE, VA.

1891

MANUFACTURINGS.

40 Roanoke Milling Company.
41 Roanoke Canning and Preserving Company
42 Mattress Factory.
43 Bridgewater Carriage Company of Roanoke.
44 Roanoke Electric Light Company
45 P. L. Terry Milling Company.
46 Roanoke Manufacturing Company.
47 Fishburns Tobacco Factory.
48 Paper Bag Factory.
49 Sash, Blind and Door Factory.
50 Roanoke Shelf Hardware Co.
51 Roanoke Blanket Mills.
52 Duval Engine Works.
53 The Bell Printing and Mfg. Co.
54 Hammond Printing Co.
55 Elevator.
56 Roanoke Black marble Co.
57 Roanoke Cold storage Co.

LAND COMPANIES.

Roanoke Development Co.
Crystal Spring Land Co.
Buena Vista Land Co.
Jeannette Land Co.
Melrose Land Co.
Belmont Land Co.
Roanoke Gas & Water Co.
The Virginia Land Co.
The Oakland Improvement Co.
Eureka Land Co.
Silvan Heights.
Wall Land Co.
Pleasant Valley Land Co.
Lansdown Land Co.
Hyde Park Land Co.
Creston Land Co.
Jefferson Land Co.
Exchange Building and Investment Co.
Inter Urban Land Co.

CHURCHES.

A Lutheran, St. Marks.
B Lutheran Chapel.
C Episcopal, St. John's.
D Catholic, St. Andrew's.
E Presbyterian, Chapel.
F A. M. E. South.
G A. High St. Baptist.
H African First Baptist.
I Calvary Baptist.
J M. E. South, Greene Memorial.
K Methodist Episcopal.
L Presbyterian.
M Christian.
N First Baptist.

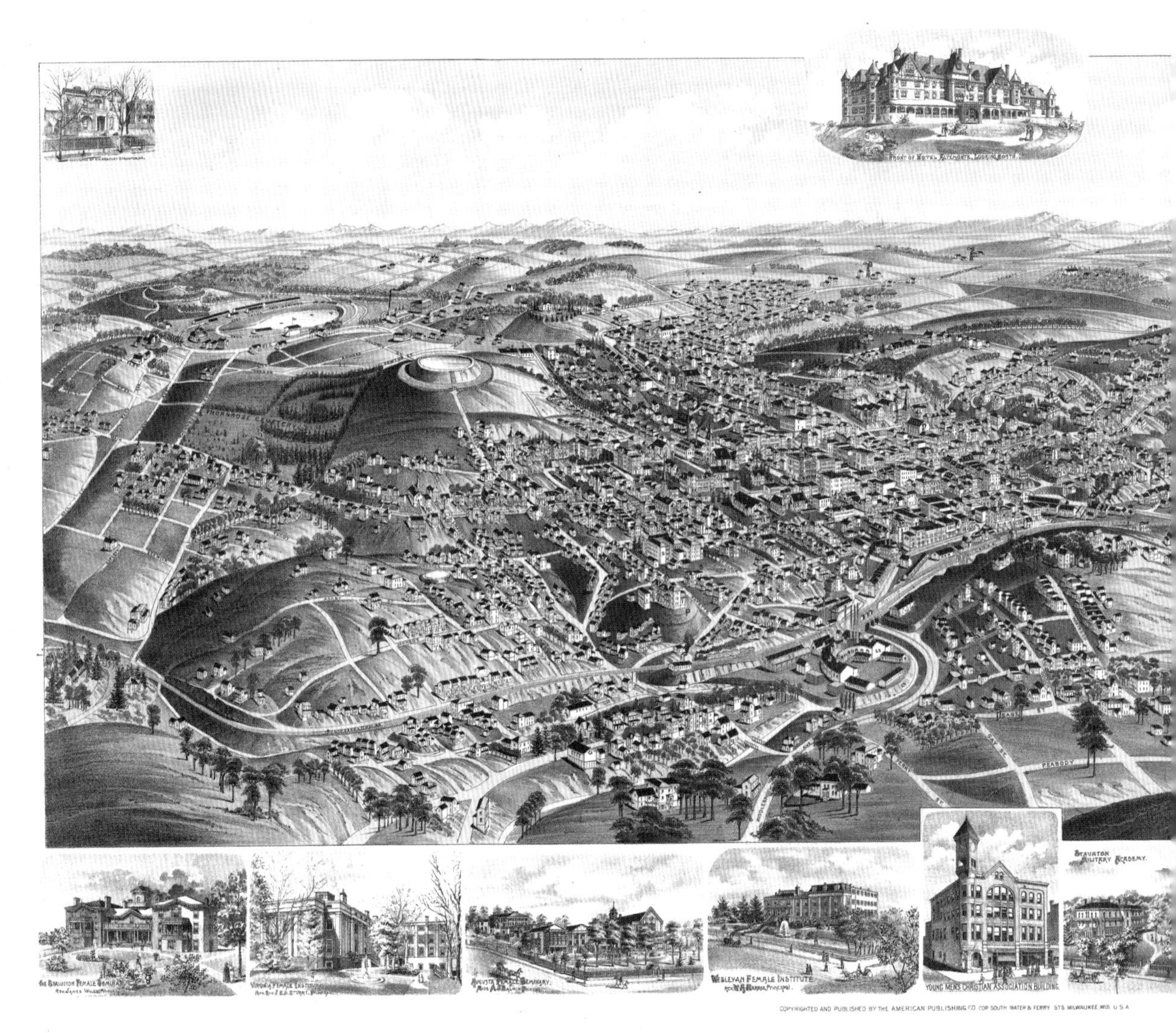

CHURCHES.

- A Trinity Episcopal Church and Chapel.
- B Baptist Church.
- C South Methodist Church.
- D Lutheran Church.
- E Second Presbyterian Church.
- G United Brethren Church.
- H First Presbyterian Church.
- K St. Francis Church.
- L Colored Churches.

SCHOOLS.

- P Staunton Female Seminary.
- Q Virginia Female Institute.
- R Augusta Female Seminary.
- S Wesleyan Female Institute.
- T Staunton Military Academy.
- U Public Schools.
- V Catholic School.
- X Colored School.

PUBLIC BUILDINGS.

1 Augusta County Court House.
2 Staunton City Hall.
3 City Water Works.
4 Fire Engine Houses.
5 Electric Light Works.
6 Gas Works.
7 Street Railway Stables.
8 Deaf and Dumb and Blind Institute of Virginia.
9 Western Lunatic Asylum of Virginia.

RAILWAY DEPOTS.

10 Chesapeake and Ohio R. R.
11 Baltimore & Ohio R. R.

PERSPECTIVE MAP OF THE CITY OF

STAUNTON, VA

COUNTY SEAT OF

AUGUSTA COUNTY.

VIRGINIA.

1891

POPULATION: 12,000.

Perspective map of the city of Staunton, Va., county seat of Augusta County, Virginia 1891. Staunton had less industry than Roanoke and seemed to put more emphasis on education, listing two female seminaries, two female institutes, Staunton Military Academy, and various public schools. It had two railroads, four hotels, two banks, and four newspapers, one named the *Vindicator.* Its diverse factories made brass, wagons, furniture, shoes, flour, fertilizer, ice, iron, and wood mantels.

Panorama of Norfolk and surroundings 1892. H. Wellge, des. Compliments of Pollard Bros. Real Estate.

The James River is in the background, with the Elizabeth River making a crescent (center) and the Southern Branch of the Elizabeth at the bottom left. The thick settlement of houses and the warehouses along the waterfront indicate that Norfolk was a thriving seaport city. An inset at the bottom right shows the harbor between Hampton and Norfolk, with lines indicating ships headed for Baltimore, Washington, Philadelphia, New York, Boston, New Orleans, Europe, and South America.

COMPLIMENTS OF

LARD BRO'S,

REAL ✻ ESTATE.

RANDOLPH MACON COLLEGE.

HOTEL BEDFORD.

BELMONT SEMINARY.

REFERENCES.

1 Knitting Mills.
2 Planing Mills.
3 Jeter Institute (Female).
4 Passenger Station N. & W. R. R.
5 Hotel Bedford.
6 Randolph—Macon Academy.
7 A. B. Claytor, Tobacco.
8 Liberty Roller Mills.
9 Episcopal Church (Colored).
10 Catholic Church.
11 Granville Sanitarium.
12 Presbyterian Church.
13 Tobacco Warehouse.
14 Jeter & McGhee, Hardware.
15 G. O. Goodwin & Co., Millinery.

REFERENCES.

16 S. B. Mosby & Co., Engineers and Contractors.
17 Baptist Church (Colored).
18 R. B. Claytor & Co., Dry Goods.
19 Stone Bros. & Co., Dry Goods.
20 Parr & Co., Real Estate.
21 W. J. Hubard, Books.
22 R. W. Coffee, Machinery.
23 Liberty Savings Bank.
24 Fisher & Hughlett Bros., Drugs.
25 Tait's Furniture Emporium.
26 A. H. Trimble, Groceries.
27 Post Office.
28 Va. Business College.
29 First National Bank.

PERSPECTIVE MAP OF

BEDFORD CITY, VA.

COUNTY SEAT OF BEDFORD CO.
1891

Perspective map of Bedford City, Va., county seat of Bedford Co. 1891.

Bedford is a small city, unlike nearby Roanoke, and in 1891 it was surrounded by tilled fields with the Peaks of Otter in the background. Features listed include one railroad, one hotel, two female seminaries, Randolph Macon Academy, and the Sanitarium. The latter was a place where tuberculosis patients could be segregated in healthful mountain air. Bedford's businesses included two livery stables; an electric company; a hardware store; a millinery; factories to make flour, cabinets, and carriages; and woolen and knitting mills.

BAPTIST CHURCH EMPORIA

METHODIST CH. BELFIELD

BIRDS EYE VIEW
OF
EMPORIA
VIRGINIA
1907

Drawn and Published by T. M. FOWLER, Morrisville, Pa.

Copyright 1907 by T. M. FOWLER, Morrisville, Pa.

BAPTIST CH. BELFIE

Birds eye view of Emporia, Virginia 1907. Drawn and published by T. M. Fowler.

Emporia is a flat city that grew along the Meherrin River at the junction of two railroads and the level route from the coast to the mountains. On this 1907 map, by T. M. Fowler, photos indicate that the town had a large dairy, a hotel, two banks, several metalworking factories, and a Masonic temple. As the county seat of Greensville County, it had a courthouse and—unlike many other Virginia areas—a large county high school.

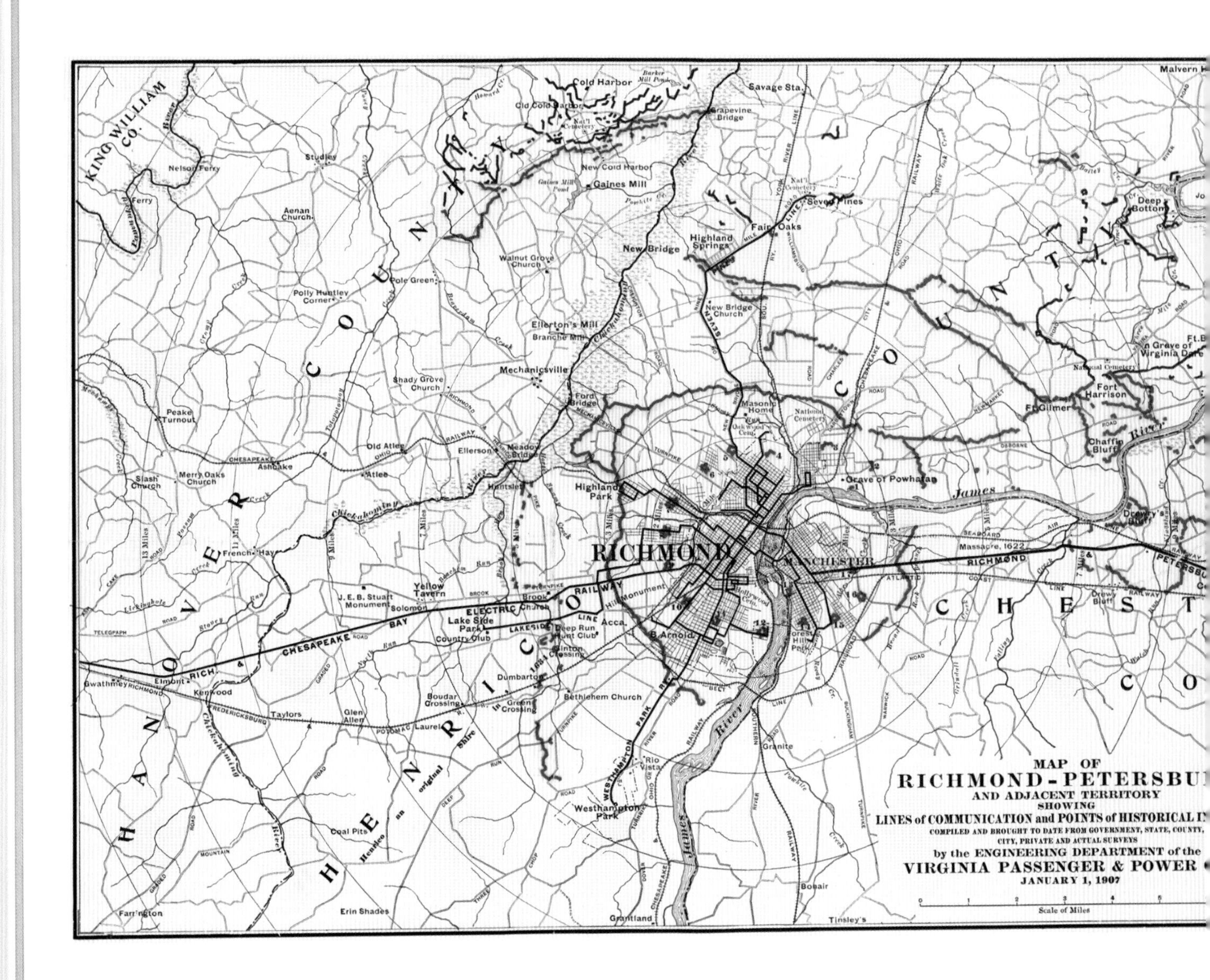

Map of Richmond-Petersburg and adjacent territory showing lines of communication and points of historical interest compiled and brought to date from government, state, county, city, private and actual surveys by the Engineering Department of the Virginia Passenger & Power Co., January 1, 1907. P. P. Pilcher, J. M. N. Allen, and J. A. B. Gibson, delineators. Calvin Whiteley Jr., C.E., Railway Dept. [Printed by] the Matthews-Northrup Works, Buffalo, N.Y. Copyright, 1903, by Virginia Passenger and Power Co. Copyright, 1907, by receivers, Virginia Passenger and Power Co.

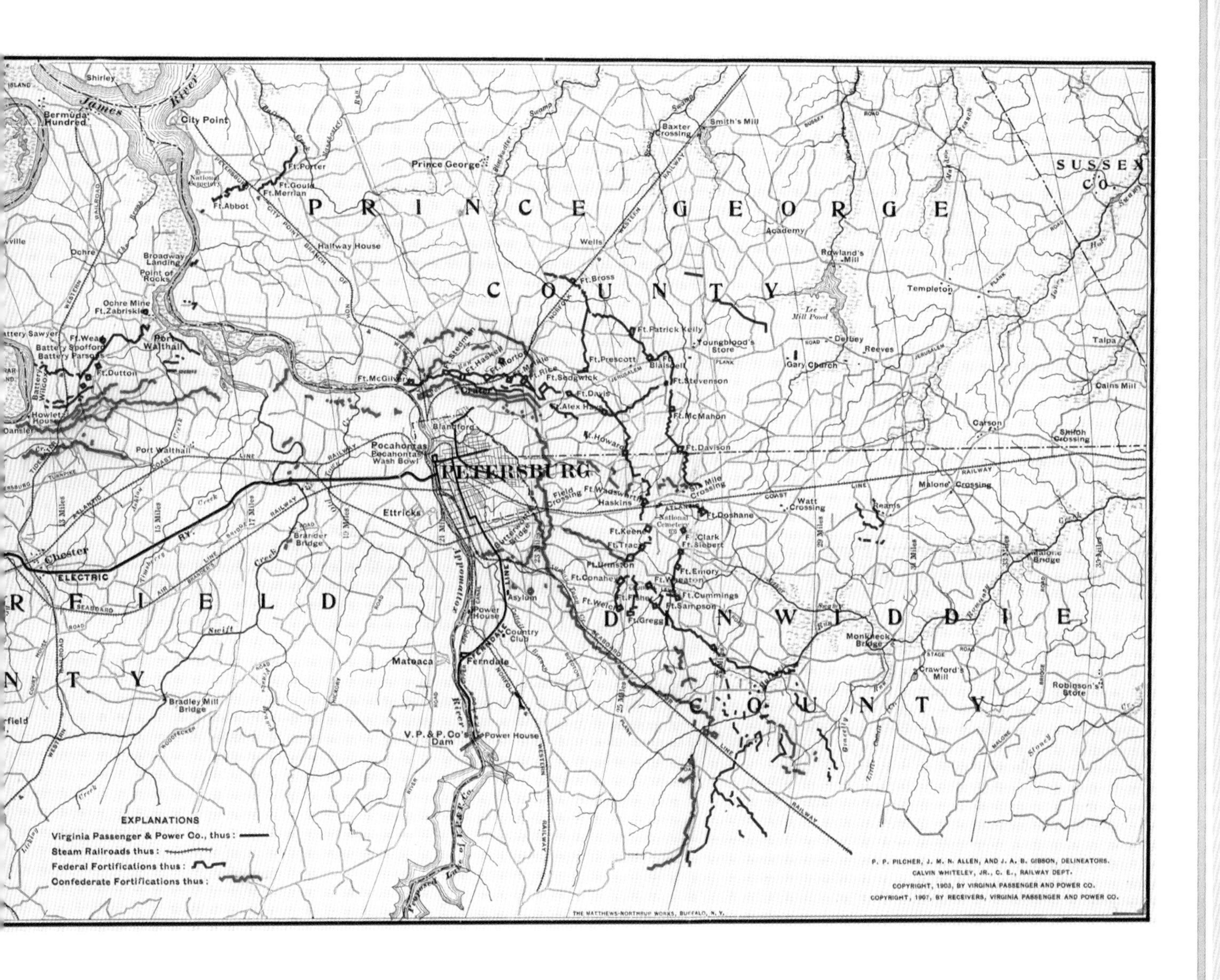

This map of the Richmond-Petersburg area, a compilation of surveys and historical data, is oriented with north to the left. Union fortifications from the Civil War are in purple, Confederate in red. Historic sites include the grave of Powhatan and Hollywood Cemetery, burial place of several famous Virginians. Mainly, though, it shows rail connections. Richmond had the first electric railways, and two are shown: Chesapeake Bay Electric and Lakeside Line. Also indicated are 7 Pines Line RR; Chesapeake and Ohio; Southern; Belt Line; and Richmond, Fredericksburg & Potomac Railroads. At that time Petersburg had the Atlantic Coastline and Seaboard Airline Railroads.

Conclusion

Each of the four centuries of Virginia's history has distinctive characteristics.

The seventeenth century was a time of fragility as the little colony clung to the edge of an unknown continent. Repeatedly, through massacre, disease, and mismanagement, the colony faced extinction, only to be saved by a new infusion of immigrants, a shipload of supplies, and a strong new leader. Toward the end of the century, conflicts with royal authorities began to emerge but were settled.

The eighteenth century was Virginia's Golden Age. It was the largest colony, both in population and land area. Virginians also led in war—first with France and then with England—and finally in the establishment of a new government, unlike any other on Earth at the time.

The nineteenth century for Virginia was a time of trial. Her size and influence were diminished and her institutions tested. The state endured a devastating war and a slow recovery. Many things would never be the same.

The dawn of the twentieth century found Virginia again at the center of power, due to its location and its military bases. Troops departed from Norfolk for the brief Spanish-American War in 1898 and again for World War I, World War II, and later military actions.

Virginians assumed positions of power and influence on the national stage. Senator Carter Glass of Lynchburg authored significant legislation, including the establishment of the Federal Reserve System. Richard Byrd explored the South Pole.

Most influential, however, was Harry F. Byrd, brother of Richard and a descendant of two of Virginia's oldest families. Joining his brother Thomas to save the family newspaper in Winchester, he became a successful apple grower and dominated the Virginia political scene for a generation in what was called the Byrd Machine. It was not the corrupt politics the term usually signifies. As governor he made Virginia a model of sound, honest government. Schools and roads were vastly improved, and the state became debt-free. After his term as governor, he became U.S. senator.

So strong was Virginia's distaste for Republican treatment during Reconstruction that most of the state voted Democratic for eighty years after the war's end. An exception was the Ninth District, in southwest Virginia, called the Fighting Ninth because there were actual shoot-outs over election disagreements. But in 1932 Virginia supported Republican Herbert Hoover and in 1952 voted as Democrats for Eisenhower. Another twenty years would pass before Republican Linwood Holton became governor.

Meanwhile, John D. Rockefeller began restoration of Colonial Williamsburg in 1926, and Shenandoah National Park was established. During the Depression, Buggs Island Dam was built, creating a huge lake and changing the topography of the state. Skyline Drive and Norfolk Botanical Garden were Works Progress Administration projects. Later, McLean House at Appomattox was restored, and in April 1950 Ulysses Grant III and Robert E. Lee IV attended a reenactment of the Civil War surrender.

World War II brought huge expansion, especially to the Hampton Roads area. The war came close to home when German subs attacked shipping off the coast of Virginia Beach, in sight of horrified sunbathers.

In the 1950s Virginia resisted integration of its public schools but settled the issue peacefully. It was the only former Confederate state in which troops were not summoned to integrate schools. Civil rights sit-ins at several Virginia locations were also peaceful. Virginia became the first state in the nation to elect a black governor, Douglas Wilder. Famous Virginians of the twentieth century include entertainers Ella Fitzgerald, Bojangles Robinson, Pearl Bailey, George C. Scott, Shirley MacLaine, Warren Beatty, and Joseph Cotten; athletes Arthur Ashe and Sam Snead; author William Styron; Gen. George S. Patton and Lewis "Chesty" Puller, most decorated Marine in history; and Edith Bolling Galt, who became the wife of President Woodrow Wilson.

In 1964 the eighteen-mile-long Chesapeake Bay Bridge-Tunnel joined Virginia's two Eastern Shore counties to the mainland. In 1952 Newport News Shipbuilding built the ocean liner USS *United States* and later built the aircraft carriers *Kennedy, Theodore Roosevelt,* and *Ronald Reagan.*

The Norfolk Naval Base is the world's largest, and across the river in Portsmouth are the Naval Hospital, the Craney Island Coast Guard Base, and the Norfolk Naval Shipyard. The National Aeronautics and Space Administration (NASA) is located across the James River, not far from the first landing of the Jamestown settlers.

Because of its proximity to Washington, D.C., Northern Virginia has seen explosive growth. Reston, the nation's first planned community, is there, as is the main airport for the nation's capital, Dulles International. Northern Virginia is also the location for the CIA and is headquarters for many high-tech firms, national charities, and lobbying organizations.

At the beginning of the twenty-first century, Bedford, Virginia, was chosen as the site of the National D-Day Memorial, in recognition of the county's military sacrifice. Bedford had America's highest percentage of casualties in the Normandy Invasion of World War II.

Politically, Virginia has flipped again. After voting for Republican presidential candidates since the late 1960s, in 2008 Virginia gave her electoral votes to Democrat Barack Obama, America's first African-American president. Yet, the state maintains its strong roots to the past. Queen Elizabeth II, monarch of our Mother Country, visited Virginia in 1957, for the Bicentennial in 1976, and again in 2007 for the four-hundredth anniversary of the Jamestown settlement.

Virginia looks back with pride on its long history and looks ahead to a bright, productive future.

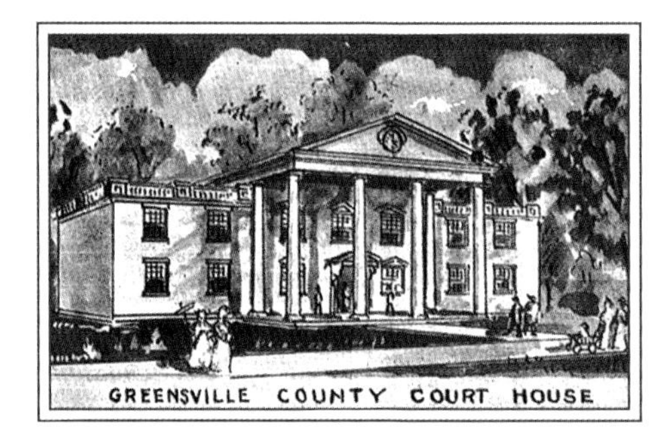

Details of map on pages 106 and 107.

Acknowledgments

The publisher and the authors gratefully acknowledge the staff at the Library of Congress for their fine work and research assistance on this book, particularly Aimee Hess, Ralph Eubanks, Ed Redmond, and Colleen Cahill.

Without the vision and professionalism of Erin Turner, this audacious project would not be the permanent achievement it is bound to be.

—Vincent Virga

Thanks to the staff of the Henderson County Public Library, and special thanks to Jerry Liedl.

—Emilee Hines

Notes

All maps come from the Library of Congress Geography and Map Division unless otherwise noted. To order reproductions of Library of Congress items, please contact the Library of Congress Photoduplication Service, Washington, D.C., 20540-4570 or (202) 707-5640.

Pages ii and 8–9 L'Isle, Guillaume de. Carta geografica dell' America settentrionale. Venezia, 1750. G3300 1750 .L4 Low 396.

Page viii Ruysch, Johann. "Universalior cogniti orbis tabula." In Claudius Ptolemeus, Geographia. Rome, 1507. G1005.1507 Vault.

Page ix Waldseemüller, Martin. "Universalis cosmographia secudum Ptholomaei traditionem et Americi Vespucii aloru[m] que lustrations," St. Dié, France?, 1507. G32001507.W3 Vault.

Page 4 White, John. Americæ pars, nunc Virginia dicta: primum ab Anglis inuenta, sumtibus Dn. Walteri Raleigh, Equestris ordinis Viri, Anno Dnī. MDLXXXV regni Vero Sereniss. nostræ Reginæ Elisabethæ XXVII, hujus vero Historia peculiari Libro descripta est, additis etiam Indigenarum Iconibus, autore Ioanne With; sculptore Theodoro de Brÿ, qui et. excud. 1590. G3880 1590 .W4 Vault.

Pages 10–11 Smith, John. Virginia, discovered and discribed by Captayn John Smith, 1606; graven by William Hole. London, 1624. G3880 1624 .S541 Vault.

Pages 12–13 Jansson, Jan. Nova Anglia, Novvm Belgivm et Virginia. 1642. G3300 1642 .J3 TIL Vault.

Pages 14–15 Herrman, Augustine. Virginia and Maryland as it is planted and inhabited this present year 1670, surveyed and exactly drawne by the only labour & endeavour of Augustin Herrman bohemiensis; W. Faithorne Sculpt. London, 1673. G3880 1670 .H4 Vault.

Pages 16–17 Ferrar, John. A mapp of Virginia discovered to ye hills, and in it's latt. from 35 deg. & 1/2 neer Florida to 41 deg. bounds of New England, Domina Virginia Farrer; John Goddard sculp. London, c1667. G3880 1667 .F3 Vault.

Page 18 Senex, John. A new map of Virginia, Mary-Land, and the improved parts of Pennsylvania & New Jersey. Rev., by I. Senex. London, 1719. G3790 1719 .S4 Vault.

Pages 22–23 Warner, John. A survey of the northern neck of Virginia, being the lands belonging to the Rt. Honourable Thomas Lord Fairfax Baron Cameron, bounded by & within the Bay of Chesapoyocke and between the rivers Rappahannock and Potowmack: With the courses of the rivers Rappahannock and Potowmack, in Virginia, as surveyed according to order in the years 1736 & 1737. 1747. G3880 1747 .W33 Vault.

Pages 24–25 Washington, George. A plan of Alexandria, now Belhaven. 1749. G3884.A3G46 1749 .W3 Vault.

Pages 26–27 Mitchell, John. A map of the British and French dominions in North America, with the roads, distances, limits, and extent of the settlements, humbly inscribed to the Right Honourable the Earl of Halifax, and the other Right Honourable the Lords Commissioners for Trade & Plantations, by their Lordships most obliged and very humble servant, Jno. Mitchell. Tho: Kitchin, sculp. London, 1755. G3300 1755 .M51 Vault.

Pages 28–29 Fry, Joshua. A map of the most inhabited part of Virginia containing the whole province of Maryland with part of Pensilvania, New Jersey and North Carolina. Drawn by Joshua Fry & Peter Jefferson in 1751. London, Thos. Jefferys, 1755. G3880 1755 .F72 Vault.

Pages 30–31 Henry, John. A new and accurate map of Virginia wherein most of the counties are laid down from actual surveys. With a concise account of the number of inhabitants, the trade, soil, and produce of that Province. London, Thos. Jefferys, 1770. G3880 1770 .H4 Vault.

Pages 32–33 Ballendine, John. A map of Potomack and James rivers in North America shewing their several communications with the navigable waters of the new province on the river Ohio. [1773?] G3880 1773 .B3 Vault.

Page 34 Washington, George. A plan of my farm on Little Huntg. Creek & Potomk. R. G. W. 1766. G3882.M7 1766 .W3 Vault.

Pages 38–39 Hutchins, Thomas. A new map of the western parts of Virginia, Pennsylvania, Maryland, and North Carolina; comprehending the River Ohio, and all the rivers, which fall into it; part of the River Mississippi, the whole of the Illinois River, Lake Erie; part of the Lakes Huron, Michigan &c. and all the country bordering on these lakes and rivers, by Thos. Hutchins, Engrav'd by T. Cheevers. London, 1778. G3707.O5 1778 .H8 Vault: Am. 6-20.

Page 40 Plan du terrein à la rive gauche de la rivière de James vis-à-vis Jamestown en Virginie ou s'est livré le combat du 6 juillet 1781 entre l'armée américaine commandée par le Mis. de La Fayette et l'armée angloise aux ordres du Lord Cornwallis. [Signé:] Desandroüins. 1781. G3884.J15S3 1781 .D4 Vault: Roch 51.

Page 41 Jefferson, Thomas. Map of the country between Albemarle Sound, and Lake Erie, comprehending the whole of Virginia, Maryland, Delaware and Pensylvania, with parts of several other of the United States of America. London, John Stockdale, 1787. G3790 1787 .J4 Vault.

Pages 42–43 Washington, George. A map of General Washington's farm of Mount Vernon from a drawing transmitted by the General. 1801. G3882.M7 1793 .W34 1801 TIL Vault.

Page 44 E. Sachse & Co. View of the University of Virginia, Charlottesville & Monticello, taken from Lewis Mountain, drawn from nature & print. in colors by E. Sachse & Co. Washington, D.C.; Richmond, Va., C. Bohn, 1856. G3884.C4:2U5A35 1856 .E2 Vault.

Pages 48–49 Böÿe, Herman. A map of the state of Virginia: reduced from the nine sheet map of the state in conformity to law, by Herman Böÿe, 1828. Virginia, 1859. G3880 1859 .B615 Vault.

Pages 50–51 Burr, David H. Map of Virginia, Maryland and Delaware exhibiting the post offices, post roads, canals, rail roads &c. by David H. Burr (late topographer to the Post Office), geographer to the House of Representatives of the U.S. London, 1839. G3709.3 1839 .B8 RR 70.

Pages 52–53 Crozet, Claudius. A map of the internal improvements of Virginia; prepared by C. Crozet, late principal engineer of Va. under a resolution of the General Assembly adopted March 15th 1848. Philadelphia, 1848. G3880 1848 .C7 RR 307.

Pages 54–55 Vaisz, W. Map of the proposed line of Rail Road connection between tide water Virginia and the Ohio River at Guyandotte, Parkersburg and Wheeling, made by W. Vaisz, top. eng. for the Board of Public Works of Virginia. Philadelphia, 1852. G3881.P3 1852 .V35 RR 81.

Page 56 Sneden, Robert Knox. Plan of Rebel redoubt and barracks [at] "Camp Misery": surveyed April 11th 1862. 1862–1865. Virginia Historical Society, P.O. Box 7311, Richmond, VA 23221-0311 USA, Mss5:7 Sn237:1, p. 99.

Pages 60–61 "Defendants Exhibit No. 14 September 9th, 1914, C.B. Filed by the State of West Virginia in the equity cause of Commonwealth of Virginia vs. State of West Virginia pending in the Supreme Court of the United States, at the hearing before Commissioner Littlefield in Richmond, Va., in August, 1914. Prepared under the direction of J. K. Anderson, Chief Engineer of the Public Service Commission of West Virginia." Map showing the location of railroads, canals, navigation projects and public institutions in which the Commonwealth of Virginia had invested money as of date January 1st. 1861: as traced from an official map in the possession of the Virginia State Library entittled [sic] "A map of the State of Virginia reduced from the Nine Sheet Map of the State in conformity to law by Herman Boye, 1828, corrected by order of the Executive 1859 by L.v. Buckholtz": together with the division line later established between Virginia and West Virginia and additional extensions made from completion of map until January 1st, 1861. [1914?] Library of Virginia, Richmond, VA 23219-8000 USA, Archives Research Services. C. 1, map accession no. 1146 (1914); c. 2 2357.

Page 62 Hausmann, A. Western Virginia from Petersburg to Warm Springs, showing the movement of the Union army, 1862, drawen [sic] by A. Hausmann. 1862. G3891.S5 1862 .H3 Vault: CW 458.

Page 63 Scene of the late naval fight and the environs of Fortress Monroe, and Norfolk and Suffolk, now threatened by General Burnside. 1862. G3882.H3S5 1862 .S3 Vault Shelf: CW 558.7.

Page 64 Map of Shenandoah County between Mt. Jackson and New Market, Virginia. 1863. G3883.S3 1863 .M3 Vault: Hotch 150.

Page 65 Magnus, Charles. Volunteer militia and eastern army guide. One hundred and fifty miles around Richmond. New York, Washington, Charles Magnus, 186-. G3884. R5S5 186- .M34 CW 632.6.

Pages 66–67 Nicholson, W. L. Map of the State of Virginia compiled from the best authorities, and printed at the Coast Survey Office. A. D. Bache, Supdt. May 1864. Compiled by W. L. Nicholson. Lith. by Chas. G. Krebs., 1864. G3880 1864 .N51 CW 493.6.

Pages 68–69 Map of Henrico County, Va.: showing fortifications around Richmond, north and east of the James River. 1864. G3883.H4S5 1864 .M3 Vault: Hotch 41.

Pages 70–71 Abbot, Henry L. Yorktown to Williamsburg. This map compiled by Capt. H. L. Abbot, Top. Eng'rs., September 1862. Photographic reduction by L. E. Walker, Treasury Department. 1862. G3883.Y6S5 1862 .A22 CW 600.

Pages 72–73 Magnus, Charles. Birds eye view of Alexandria, Va. New York; Washington, D.C., Chas. Magnus, 1863. G3884.A3A3 1863 .M32 Vault: CW 522.3.

Pages 74–75 Henderson & Co. Map of Appomattox Court House and vicinity. Showing the relative positions of the Confederate and Federal Armies at the time of General R. E. Lee's surrender, April 9th 1865. Baltimore, A. Hoen & Co., c1866. G3884.A6S5 1865 .H3 CW 524.

Page 76 Hotchkiss, Jedediah. The Am. Manufacturer's map of the New River & the Flat-top coking coal fields of the Virginias, by Jed. Hotchkiss, Cons. M.E., Staunton, Va., Sept. 1886. G3893.F3H9 1886 .H6 Vault: Hotch 244.

Pages 80–81 Hotchkiss, Jedediah. Hotchkiss' geological map of Virginia and West Virginia, the geology by Prof. William B. Rogers, chiefly from the Virginia State survey, 1835–41, with later observations in some parts. Richmond, Va., A. Hoen & Co., 1875. G3881.C5 1875 .H6 Vault: Hotch 219.

Pages 82–83 Maury, Matthew Fontaine. Map showing the economic minerals along the route of the Chesapeake & Ohio Rail Way to accompany the geological report of Thomas S. Ridgway. 1872. G3881.H1 1872 .M38 RR 365.

Page 84 Hotchkiss, Jedediah. Map of Piedmont Virginia, by Jed. Hotchkiss, Top. Eng. Staunton, Va. 1873; D. C. Humphreys del. G3880 1873 .H6 Vault: Hotch 210.

Page 85 Hotchkiss, Jedediah. Map of middle Virginia, by Jed. Hotchkiss, Top. Eng., Staunton, Va., 1873. G3880 1873 .H62 Vault: Hotch 212.

Pages 86–87 Hotchkiss, Jedediah. Geological map of the Potomac basin west of Blue Ridge, Virginia and West Virginia: showing the relations of its upper Potomac coal basin and the iron-ore bearing areas in reference to the West Virginia Central and Pittsburg [sic] R.R., and its connections and extensions, compiled from surveys by W. B. Rogers and others by Jed. Hotchkiss, Consulting Eng., etc. 1882. G3884.B68C5 1882 .H6 Vault: Hotch 220.

Pages 88–89 Rogers, William Barton. Geological map of Virginia & West Virginia showing their chief geological sub-divisions, by Prof. William B. Rogers on basis of the physical & pol. map of A. Guyot. [187?] G3881.C5 187- .R6 Vault: Hotch 218.

Pages 90–91 Hotchkiss, Jedediah. Map of part of the great Flat-top coal-field of Va. & W. Va. showing location of Pocahontas & Bluestone collieries, May 1886. Staunton, Va., Eng. Office of Jed. Hotchkiss, 1886. G3883.T3H9 1886 .H6 Vault: Hotch 238.

Pages 92–93 Hotchkiss, Jedediah. A centennial map of the Lexington Presbytery of the Synod of Virginia of the Presbyterian Church in the United States, organized Sept. 26, 1786, centennial Sept. 26, 1886, at Timber Ridge Church, by Jed. Hotchkiss, Top. Eng. 1886. G3880 1886 .H6 Vault: Hotch 214.

Page 94 Carter, William T. Virginia, Fairfax-Alexandria counties sheet, soil map; soils surveyed by Wm. T. Carter Jr., in charge, and C. K. Yingling Jr. Washington, D.C., U.S. Dept. of Agriculture, Bureau of Soils, 1915. G3883.F2J3 1915 .C3.

Pages 98–99 Perspective map of the city of Roanoke, Va. 1891. Milwaukee, American Publishing Co., 1891. G3884. R6A3 1891 .A6.

Pages 100–101 Perspective map of the city of Staunton, Va., county seat of Augusta County, Virginia 1891. Milwaukee, American Publishing Co., 1891. G3884.S8A3 1891 .A6.

Pages 102–103 Wellge, Henry. Panorama of Norfolk and surroundings 1892. H. Wellge, des. Compliments of Pollard Bros. Real Estate. Milwaukee, American Publishing Co., 1892. G3884.N6A3 1892 .W41.

Pages 104–105 Wellge, Henry. Perspective map of Bedford City, Va., county seat of Bedford Co. 1891. Milwaukee, American Publishing Co., 1891. G3884.B3A3 1891 .W4.

Pages 106–107 and 112 Fowler, T. M. Birds eye view of Emporia, Virginia 1907. Morrisville, Pa., T. M. Fowler, 1907. G3884.E5A3 1907 .F6.

Pages 108–109 Whiteley, Calvin Jr., C.E., Railway Dept. Map of Richmond-Petersburg and adjacent territory showing lines of communication and points of historical interest compiled and brought to date from government, state, county, city, private and actual surveys by the Engineering Department of the Virginia Passenger & Power Co., January 1, 1907. P. P. Pilcher, J. M. N. Allen, and J. A. B. Gibson, delineators. Virginia Passenger and Power Co., 1907. G3884.R5 1907 .V5 CW 647.

About the Authors

VINCENT VIRGA earned critical praise for *Cartographia: Mapping Civilization* and coauthored *Eyes of the Nation: A Visual History of the United States* with the Library of Congress and Alan Brinkley. Among his other books are *The Eighties: Images of America,* with a foreword by Richard Rhodes; *Eisenhower: A Centennial Life,* with text by Michael Beschloss; and *The American Civil War: 365 Days,* with Gary Gallagher and Margaret Wagner. He has been hailed as "America's foremost picture editor" for having researched, edited, and designed nearly 150 picture sections in books by authors including John Wayne, Jane Fonda, Arianna Huffington, Walter Cronkite, Hillary Clinton, and Bill Clinton. Virga edited *Forcing Nature: Trees in Los Angeles,* photographs by George Haas for Vincent Virga Editions. He is the author of six novels, including *Gaywyck, Vadriel Vail,* and *A Comfortable Corner,* as well as publisher of ViVa Editions. He has a Web site through the Author's Guild at www.vincentvirga.com.

EMILEE HINES is a native Virginian with a love for the state and an interest in its history. A graduate of Lynchburg College with a master's degree from the University of North Carolina at Chapel Hill, she has taught in Virginia and Kenya. She's been fascinated with maps since sitting in a one-room school as a first-grader, studying maps posted around the room, and later in Latin and Spanish classes, vowing to see the countries that speak Romance languages.

Hines has coauthored nine volumes of *Old Virginia Houses* and is the author of *It Happened in Virginia* (Globe Pequot Press) and *More than Petticoats: Remarkable Virginia Women* (Globe Pequot Press); a memoir, *East African Odyssey;* a droll novel, *Burnt Station;* and more than three hundred published articles and short stories. She was married to the late Thomas B. Cantieri and has a daughter, Catherine, who lives in Portsmouth, Virginia. You can learn more about Hines at www.emileehines.com.